Exploring *The*
BUILDING BLOCKS
of
Science

Book 7

LABORATORY NOTEBOOK

REBECCA W. KELLER, PhD

Illustrations: Janet Moneymaker

Exploring the Building Blocks of Science Book 7 Laboratory Notebook
ISBN 978-1-941181-18-8

Published by Gravitas Publications Inc.
Real Science-4-Kids®
www.realscience4kids.com
www.gravitaspublications.com

Contents

Experiment 1

Using Science

Introduction

Explore how science is used every day.

I. Think About It

❶   How many times a day do you think you use science? When and how do you use it?

cars, food, water, TV, teeth brushing, Shower, doors, Phones, School, chlothes, light swiches, and glasses

❷   What products do you use that are a result of science?

a water filter cleans your water. People have to build doors with tools and medel, wood, and rubber. bakeing with ingridients people find and/or make withe can result with a yummy meal. Plastic and glass isle good start for glasses

❸   What three items can you think of that were discovered/invented using science that have improved people's lives?

Thomas Eddison invented The lightbulb and now, people don't have to aiway light a fire for light. docters have made many cures for bad and miserable sicknesses. They have saved lives. Phones were invented and now People can call in emergencies and help will be on the way.

❹ What three items can you think of that were discovered/invented using science that have harmed people's lives?

covid 19 has killed many lives. cars have been helpful, but be getin accidents too. drugs have seperated families and killed thousands. (gun killed and kill loads)

❺ What three items do you use every day that require electricity or gasoline to run? What scientific discoveries had to be made before these items could be produced?

The cars use gasoline but they needed to make sure they were safe before useing them. the oven is very dangerous. They needed a way to make it safer. the Lawnmower needed to know how to work it. how dd you put in The gas and what kind of levy does it hed

❻ If you were a scientist with your own research lab, what three items or discoveries would you most like to make? Why?

Video games. Because I would enjoy choosing how they pray, building the map and what they do! The Phone. I think it would be an awesome Project because I've always wanted to know whats inside of a Phone and how to build it. a car. I could desighn it, and See the insides of it.

II. Experiment 1: Using Science

Date _____

Objective _____

_____

Materials

internet or library
notebook or blank paper
imagination

EXPERIMENT

❶ Review the items you listed in *Questions 3-5* in the *Think About It* section. Choose one item you would like to know more about. Describe the scientific knowledge needed to make it. What materials were used? How are the materials made or obtained? What scientific knowledge was needed? What can you find out about the manufacturing process and how is science involved in this? Be as specific as you can. Use the internet or library to look up how the item is made and what materials it is made from. For example, if you have chosen a battery operated or electric toothbrush, what type of plastic is it made of? What are the bristles made of? How does it get its power? How was it manufactured? Research the item and take notes in a notebook or on blank paper. In the *Results* section, summarize the information.

❷ Pick one of the items or discoveries you listed *Question 6*. Think about how you might take your item to its full development or how your discovery could be developed into a product. For example, if you are discovering a cure for a disease, think about what you would need to do to go from the idea to seeing your cure saving lives. Research how drugs are developed, who develops them, how companies make them, how doctors find out about them, and finally, how the drugs would be given to people. If you are creating a new product, such as a flying car or a tent that captures water from the air, think about what steps you would need to take to go from the idea to seeing the product manufactured and sold. Do research to discover details such as how to make a prototype or test product, how to do market research for your product, how to take your product to a manufacturer, what needs to be set up to build and produce your product, and finally, how you could distribute your product.

As you do research, take notes in a notebook or on blank paper. When you are finished, summarize the information in the *Results* section.

Results

Summarize the results of your research for Steps ❶-❷. If needed, use extra paper and fasten it in this section.

_____

_____

_____

_____

_____

_____

_____

_____

_____

_____

_____

_____

_____

_____

_____

_____

III. Conclusions

Discuss what you learned as you researched how science is used in your everyday life and what you learned about how you might create a product or a cure that could help others.

_____

_____

_____

_____

_____

_____

_____

_____

_____

_____

_____

_____

_____

_____

_____

IV. Why?

Write your own *Why!* To write this section, first think of questions you want to address, such as: Why is it important to look at how science is used in our everyday lives? Why do scientists need to understand not only how to create new products but how they might be used in ways that could either help or harm people, other living things, or our planet? Why is research important? Also think about the "hows" that go with the why questions.

_____

_____

_____

_____

_____

_____

_____

_____

_____

_____

_____

_____

_____

V. Just For Fun

What product can you think of that you would like to create if you had your own lab? It might be an improvement on an existing type of product or an idea that you have for something new. To help come up with a product, review your answers in the *Think About It* section, look at *Chapter 1* in the *Student Textbook*, recall something of interest you noticed while doing the research for the *Using Science* experiment, etc. Make notes that summarize how you would research and make the product, then make drawings and/or a small model.

New Product Idea!

New Product Idea—More Details!

Experiment 2

Mix It Up!

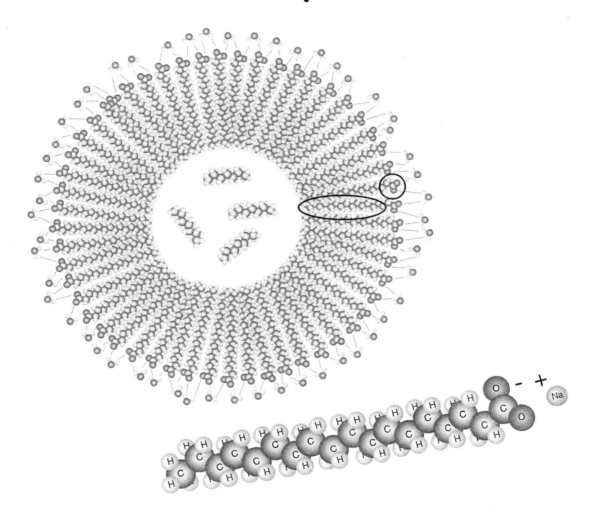

Introduction

How well do substances mix? Find out!

I. Think About It

CHEMISTRY

❶ Have you noticed two substances that will mix? What are they and how do they mix?

_____

_____

_____

_____

❷ Have you noticed two substances that will not mix? What are they and what happens when you try to mix them?

_____

_____

_____

_____

❸ If two substances mix, do you think it will be easy to separate them? Why or why not?

_____

_____

_____

_____

❹ Do you think any two solutions you find in your house will mix together? Why or why not?

_____

_____

_____

_____

_____

❺ Do you think some substances will dissolve in a solution and some will not? Why or why not?

_____

_____

_____

_____

_____

❻ Do you think you might be able to change how two substances mix? If so, how would you do it?

_____

_____

_____

_____

_____

II. Experiment 2: Mix It Up!   Date _____

Objective   _____

_____

Hypothesis   _____

_____

Materials

water   food coloring
ammonia   dish soap
vegetable oil   eyedropper
rubbing alcohol   measuring cup and measuring spoons
melted butter   marking pen
vinegar   spoon
small jars (7 or more)

EXPERIMENT

Part I: What Mixes?

❶   Use the grid in the *Results* section to record your observations. Label both the top and one side with names of the following liquids: water, ammonia, vegetable oil, rubbing alcohol, melted butter, and vinegar. You will be mixing these substances together. Determine how to avoid making the same mixture twice and mark the grid accordingly.

❷   Take 6 small jars and put 60 milliliters (1/4 cup) of each liquid in its own jar and label each jar.

❸   Add a drop of food coloring to each jar.

❹   Using a clean jar each time, stir together 15 ml (1 tablespoon) of a liquid that has not been colored and 15 ml (1 tablespoon) of a colored liquid. Before mixing the two, make a prediction about whether they will mix. In the white boxes on the next page, record whether or not the two liquids mix. Was your prediction correct?

❺   Repeat Step ❹ for each combination of liquids.

Results

	Water	ammonia	Vegetable oil	rubbing acohol	melted butter	vigar
water						
Ammonia						
vegetable oil						
rubbing Acohol						
melted butter						

Part II: Soap, Oil, and Water

❶   Measure 60 ml (1/4 cup) of water into a small glass jar. Add one drop of food coloring.

❷   Add 15 ml (1 tablespoon) of vegetable oil to the water.

❸   Stir the water and oil. Record your results below.

❹   Add 15 ml (1 tablespoon) of liquid dish soap to the oil/water mixture.

❺   Stir thoroughly. Record your results.

❻   Add another 15 ml (1 tablespoon) of liquid dish soap to the mixture, and stir thoroughly. Record your results.

Results

Oil + water

_____

_____

_____

_____

Oil + water + 15 ml (1 tablespoon) soap

_____

_____

_____

_____

Oil + water + 30 ml (2 tablespoons) soap

_____

_____

_____

_____

III. Conclusions

What conclusions can you draw from your observations?

_____

_____

_____

_____

_____

_____

_____

_____

_____

_____

_____

_____

_____

_____

_____

_____

_____

_____

CHEMISTRY

CHEMISTRY

IV. Why?

Most of the things we encounter in daily life are mixtures rather than pure substances. In this experiment you explored homogeneous and heterogeneous mixtures.

Homogeneous mixtures are mixtures in which the molecules of the different substances are evenly distributed throughout the mixture. Salt water is an example of a homogeneous mixture. Other examples include: alcohol/water mixtures, sugar/water mixtures, vinegar, and household ammonia.

Heterogeneous mixtures are those mixtures where the molecules are not evenly distributed throughout the mixture. Ice water is an example of a heterogeneous mixture. Other examples include: sand, concrete, ice cream floats, and salad dressing.

An example of a mixture that looks homogeneous but is actually heterogeneous is milk. Milk is a colloid. A colloid has very small molecules suspended in it that are not evenly distributed in the solution and are too small to see with our eyes. All colloids are heterogeneous and cloudy. True homogeneous solutions are clear or colored but not cloudy.

In this experiment you observed the rule "like dissolves like" which states that substances that are alike will dissolve in one another, and substances that are not alike will not dissolve in one another. "Like" in this context means that both substances are either polar (charged) or both are nonpolar (uncharged). A polar molecule is one that has both a positive (+) end and a negative (-) end. Polar simply means having two opposite directions or natures. In the case of molecules, polar means that there are two oppositely charged ends on the same molecule. "Unlike" means that one substance is made of polar, or charged, molecules and the other is made of nonpolar, or uncharged, molecules.

Water is very polar. In this experiment you observed that other molecules that are polar will dissolve in water. All OH ends are also very polar wherever they occur on a molecule. Molecules with these OH ends will easily mix with water. Acetic acid (vinegar) and sugar molecules contain polar OH groups, and both easily dissolve in water.

You also observed that nonpolar molecules, such as vegetable oil, will not dissolve in water. However, nonpolar molecules will mix with other nonpolar molecules.

Soap molecules have both polar ends and nonpolar ends. Soap "allows" oils to dissolve in water. It does this by forming tiny droplets called micelles that are suspended in the surrounding water. These little oil droplets can then be washed away by the excess water.

V. Just For Fun

Black Is Black?

Date _____

Objective _____

_____

Hypothesis _____

_____

Materials

ballpoint ink pens of various
    colors, including black
rubbing alcohol
coffee filters (white)
several small jars

cardboard shoebox (or similar size box)
tape
measuring cup
scissors
ruler

EXPERIMENT

❶ Pour 60 milliliters (1/4 cup) of alcohol into each of several small jars—one jar for each color ink you will be testing. Label each jar with a color of ink.

❷ Remove the thin plastic ink tube from inside each ballpoint pen.

❸ Pull off the top or cut off the end of each ink tube.

❹ Take an ink tube and swirl the open end of it briefly in the alcohol. Make sure that some of the ink gets dissolved but don't let the alcohol get too dark in color. Repeat for each ink tube using the jar labeled for that tube.

❺ Cut the coffee filter paper into thin strips 6-12 millimeters (1/4 to 1/2 inch) wide and 13-15.5 centimeters (5 to 6 inches) long.

❻ In the experimental setup, one end of each paper strip will be placed in the dissolved ink in a jar and the alcohol allowed to migrate upwards. It is best if you can suspend the strips in the alcohol without letting them touch the sides of the jars because the alcohol won't migrate past this point. To set up the experiment, tape the paper strips to the top edge of a cardboard box that is on its side and suspend the strips in the glass jars. Or you may come up with a different idea for an experimental setup.

❼   The colors in the ink will migrate up the absorbent paper strips. Let the strips sit in the alcohol overnight.

Results

Take the paper strips out of the alcohol and let them dry, then tape them in the following chart. Write down each original ink color and record the different colors each is made of.

Orange Ink: made of yellow and red						

Unknown Ink

Repeat the previous steps using one or more samples of an unknown ink. Try to determine the color of the unknown ink sample by comparing your results with those of the previous samples.

Unknown	Color of the unknown ink:
	_____

Unknown	Color of the unknown ink:
	_____

Unknown	Color of the unknown ink:
	_____

Unknown	Color of the unknown ink:
	_____

Unknown	Color of the unknown ink:
	_____

Unknown	Color of the unknown ink:
	_____

CHEMISTRY

III. Conclusions

What conclusions can you draw from your observations?

_____

_____

_____

_____

_____

_____

_____

_____

_____

_____

_____

_____

_____

_____

Experiment 3

Show Me the Starch!

CHEMISTRY

Introduction

Which food items do you think contain starch? Can you find out?

I. Think About It

❶ Do you think potatoes contain starch? Why or why not?

_____

_____

_____

_____

❷ What type of food would you eat if you wanted lots of energy to climb a mountain or run a marathon? Why?

_____

_____

_____

_____

❸ Do you think there is a way to test for starch? Why or why not?

_____

_____

_____

_____

❹ What do you think happens when a banana or other fruit ripens?

_____

_____

_____

_____

_____

❺ Do you think you could detect the changes in carbohydrates that a banana undergoes as it ripens? Why or why not?

_____

_____

_____

_____

_____

❻ What do you think is different about the structure of starch and the structure of sugars?

_____

_____

_____

_____

_____

_____

II. Experiment 3: Show Me the Starch!   Date _____

Objective   _____

_____

Hypothesis   _____

_____

Materials

tincture of iodine [Iodine is VERY poisonous — DO NOT EAT any food items with
iodine on them.]

a variety of raw foods, including:

pasta

bread

celery, sliced

potato, sliced

banana and other fruits

liquid laundry starch (or a borax and corn starch mixture)

absorbent white paper

eyedropper

cookie sheet

marking pen

EXPERIMENT

❶   Take several food items and a piece of absorbent paper and place them on a cookie
sheet.

❷   Label the piece of absorbent paper **Control.** Using the eyedropper, put a small amount
of liquid starch (or a borax and corn starch mixture) on the paper. Let it dry.

❸   Add a drop of iodine to the control starch on the absorbent paper. In the chart in the
Results section, record the color.

❹   Add iodine to each of the food items and record the color for each.

❺   Compare the color of the "control" to the color of each food item.

Results

Food Item	Color
Control	

III. Conclusions

What conclusions can you draw from your observations?

_____

_____

_____

_____

_____

_____

_____

_____

_____

_____

_____

_____

_____

_____

_____

_____

IV. Why?

Natural starches are a mixture of 10-20% amylose and 80-90% amylopectin. Both amylose and amylopectin contain chains of linked glucose monomers. Amylopectin forms a branched structure and amylose forms a helical coil.

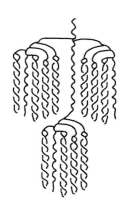

Amylopectin forms a branched structure

Courtesy of Laghi.l,
CC BY SA 3.0

Amylose forms a helical structure

The amylose in starchy foods is responsible for the formation of the deep blue color in the presence of iodine. The iodine molecule inserts itself inside the amylose coil, forming an iodine-amylose complex. A complex is a compound, or joint, molecule made of more than one molecule that are not chemically bonded.

In this experiment you used the iodine-starch test to detect the presence of complex carbohydrates. Complex carbohydrate molecules have chains of a few to thousands of monosaccharide units. Specifically, you tested for the presence of amylose. When iodine is added to amylopectin, cellulose, or small sugars, iodine will keep its orange or yellow color, but when amylose is present, a blue-black color change occurs. The complex creates the blue color by bending light waves in a way that causes wavelengths that create a deep blue to be reflected. Alone, iodine or amylose do not bend light in this way, but when combined in a complex, they do.

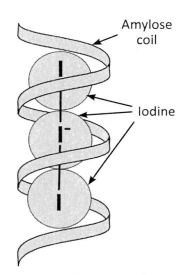

Iodine-amylose complex

Courtesy of Roland Mattern,
CC BY SA 3.0

V. Just For Fun

A Banana Ripens!

Test what happens when a banana ripens.

❶ Cut a slice off a green banana and test for starch using the iodine-starch test from the first experiment. Leave the banana at room temperature for the entire experiment.

❷ In the following chart, record your observations.

❸ Test a new slice of banana each day for several days as the banana ripens. Record your observations each time.

Day 0	Day 1	Day 2	Day 3	Day 4	Day 5

❹ Review your data. What conclusions can you draw as a result of your experiment?

_____

_____

_____

_____

_____

_____

CHEMISTRY

Experiment 4

Gooey Glue

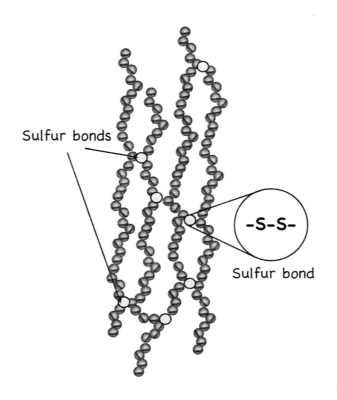

Sulfur bonds

-S-S-

Sulfur bond

CHEMISTRY

Introduction

Can a polymer change its properties? Find out in this experiment.

I. Think About It

❶ How would you define a polymer?

_____

_____

_____

_____

_____

❷ Where do you think you would find polymers?

_____

_____

_____

_____

_____

❸ Do you think polymers are important to your everyday life? Why or why not?

_____

_____

_____

_____

_____

❹ Do you think polymers always stay the same, or do you think their properties can be changed? Why?

_____

_____

_____

_____

_____

❺ Do you think it could be useful if the properties of polymers could be changed? Why?

_____

_____

_____

_____

_____

❻ What do you think life would be like if there were no polymers? Why?

_____

_____

_____

_____

_____

CHEMISTRY

CHEMISTRY

II. Experiment 4: Gooey Glue

Date _____

Objective _____

_____

Hypothesis _____

_____

_____

Materials

liquid laundry starch (or a mixture of borax, corn starch, and water)
Elmer's white glue
Elmer's blue glue (or another glue different from white glue)
water
2 small jars
marking pen
Popsicle sticks for stirring
measuring cup

EXPERIMENT

Part I

❶ Open the bottle of Elmer's white glue. Put a small amount on your fingertips. Note the color and consistency (sticky, dry, hard, soft) of the glue. Record your observations in the *Results* section.

❷ Now look carefully at the liquid starch. Pour a small amount on your fingers or in a jar. Note the color and consistency of the starch. Record your observations.

❸ Take one of the jars and put 60 ml (1/4 cup) of water into it.

❹ Note the level of water in the jar and use a marker to draw a small line at the water level.

❺ Add another 60 ml (1/4 cup) of water and use a marker to draw a small line at the water level.

❻ Pour the water out.

❼   Fill the jar to the first mark with Elmer's glue.

❽   Fill the jar to the second mark with liquid starch.

❾   Mix the glue and the starch with a Popsicle stick. Record any changes in consistency and color.

❿   Take the mixture out of the jar, and knead it with your fingers. Observe the consistency and color. Record your results.

Results

Observations for Elmer's white glue:

_____

_____

_____

_____

Observations for liquid starch:

_____

_____

_____

_____

Observations for the mixture of Elmer's white glue and equal amount of liquid starch:

_____

_____

_____

_____

_____

_____

CHEMISTRY

Part II

❶   Take another jar, and follow Steps ❸-❻ in **Part I** of this experiment.

❷   Fill the jar to the first mark with Elmer's blue glue or another glue that is different from the white glue.

❸   Add liquid starch to the second mark.

❹   Mix.

❺   Record your observations.

Results

Observations for mixture of blue glue and liquid starch:

_____

_____

_____

_____

_____

_____

_____

_____

_____

_____

_____

_____

_____

_____

III. Conclusions

What conclusions can you draw from your observations?

_____

_____

_____

_____

_____

_____

_____

_____

_____

_____

_____

_____

_____

_____

_____

_____

CHEMISTRY

IV. Why?

Polymers are long chains of molecules that are in repeating units. There are many different kinds of polymers, both natural and synthetic. These include all plastics, foods, clothing, and things made of wood. Even our bodies contain polymers. The structure of polymers helps determine their physical properties. For example, cellulose and starch are made of exactly the same molecules, but they are put together differently. This gives cellulose and starch very different properties.

Because they have different properties, polymers are very useful for a wide variety of purposes; for example, a plastic can be soft, hard, or stretchy. Differences in properties are due to differences in the polymer structure whether the polymer is naturally occurring or synthetic. For example, plastics made of linear chains of polyethylene ("linear" refers to those without side branches) are very different from chains that contain side branches. The molecules in linear chains can pack closely, which makes the plastic hard and stiff. Branched molecules, on the other hand, cannot pack together tightly, and this makes the plastic soft.

Polymer properties can also change by simply hooking the individual polymer chains together which is what you observed in this experiment. Elmer's white glue is a polymer made of single chains of molecules that slide past each other and give the glue its slippery, sticky property. The liquid laundry starch (or borax-cornstarch) is not a polymer but individual molecules. When you mixed the liquid laundry starch or the cornstarch-borax mixture into the glue, a chemical reaction occurred that caused molecules of laundry starch (or cornstarch-borax) to form bonds between chains of the glue polymer. The bond formed between two single polymer chains is called a cross-link. Once the polymer chains in the glue were cross-linked, they could no longer slide past each other.

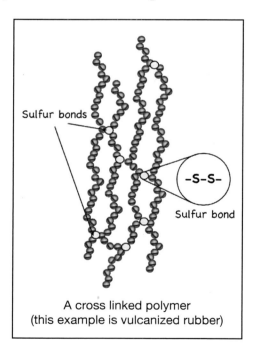

Sulfur bonds

−S−S−

Sulfur bond

A cross linked polymer
(this example is vulcanized rubber)

V. Just For Fun

Make nylon!

❶   Look up the chemistry for Nylon 6-10 on the internet or at the library. In the following space, record what you discover.

_____

_____

_____

_____

_____

❷   Get out the nylon synthesis kit, don your safety gear, and follow the instructions that come with the kit. Record your observations and conclusions in the space below.

_____

_____

_____

_____

_____

_____

_____

_____

_____

_____

_____

Experiment 5

Amylase Action

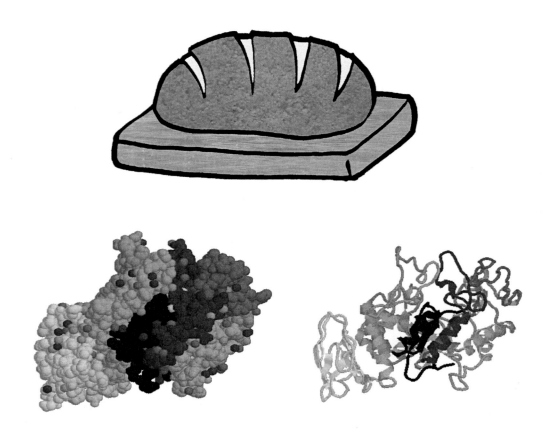

Introduction

Find out what happens when you chew bread.

I. Think About It

❶   Do you think proteins are important to life? Why or why not?

_____

_____

_____

_____

❷   If you were a protein machine, what kind of job might you be doing and how would this job help the body function?

_____

_____

_____

_____

❸   What do you think happens when you put food in your mouth and chew it? Why?

_____

_____

_____

_____

❹   Do you think you need proteins to digest food? Why or why not?

_____

_____

_____

_____

_____

_____

❺   Do you think proteins in the body can be modified? Why or why not?

_____

_____

_____

_____

_____

_____

❻   How do you think DNA is structured? Why do you think it is important for life?

_____

_____

_____

_____

_____

_____

_____

II. Experiment 5: Amylase Action   Date _____

Objective _____

_____

Hypothesis _____

_____

_____

Materials

tincture of iodine [VERY POISONOUS—**DO NOT EAT** any food items that have
  iodine on them]
bread
timer
wax paper
marking pen
cup

EXPERIMENT

❶ Break the bread into several small pieces.

❷ Chew one piece for 30 seconds (use the timer), chew another piece for 1 minute, and a
   third piece for as long as possible (several minutes).

❸ Each time, after chewing the bread, spit it onto a piece of wax paper. Using the marking
   pen, label the wax paper with the length of time the bread has been chewed.

❹ Take three small pieces of unchewed bread, and place one next to each of the chewed
   pieces.

❺ Add a drop of iodine to each piece of bread — chewed and unchewed.

❻ Record your observations in the chart in the *Results* section.

❼ Take two more pieces of bread. Collect as much saliva from your mouth as you can
   (spit into a cup several times). Soak both pieces of bread in the saliva. Place one piece

in the refrigerator, and leave the other piece at room temperature. Let them soak for 30 minutes.

❽ After 30 minutes add a drop of iodine to each. Record your results.

Results

Chewed Bread

30 seconds

1 minute

Several minutes

Unchewed bread

Bread + Saliva — 30 Minutes

Refrigerated

Not refrigerated

III. Conclusions

What conclusions can you draw from your observations?

_____

_____

_____

_____

_____

_____

_____

_____

_____

_____

_____

_____

_____

_____

_____

_____

_____

_____

_____

CHEMISTRY

IV. Why?

Amylose is a starch found in bread and potatoes. In our saliva, we have a protein machine called amylase that cuts the amylose (starch) polymer chain into the sugar molecules it is made of. Iodine reacts with starch but not with sugars. In this experiment, as a piece of bread was chewed, the starch (amylose) was mixed with the amylase in the saliva. The more the bread was chewed, the more starch was broken down into sugar and the less starch was available to react with the iodine.

Amylase is an enzyme, and enzymes are highly specific. That is, they work only on certain molecules. For example, amylase cannot digest cellulose, it can only digest amylose.

The names of amylose and amylase are very similar. Many enzymes (protein machines) are named after the molecules they work on. Amylase is the enzyme that breaks down amylose, the starch. Notice that the endings differ: "-ase" for the enzyme, "-ose" for the starch. Cellulase, a different enzyme, digests cellulose. The name of an enzyme usually tells which molecule the enzyme works on.

No matter how complicated the shape of a protein may look from the outside, it is just one long, folded-up chain on the inside. In this illustration, the picture on the left shows how human salivary amylase looks when all the atoms are shown. It is very compact, with little or no space between the atoms. The picture on the right is a simplified view that shows only the path of the protein chain without all the atoms. It's still complicated, but by looking carefully one can see that there is only one continuous chain. Also notice the coils. These are a common feature of almost all proteins. Even though the folded protein may look like a shapeless blob, it is actually folded very carefully to get the shape needed to carry out its function.

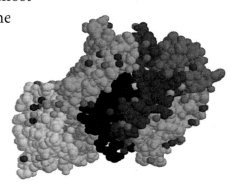
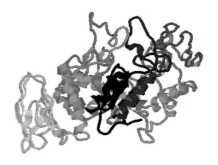

Scientists still do not know everything about all of the proteins inside cells and the roles they play. Many protein assemblies are very complex. Protein machines are very sophisticated motors, rotors, gears, pumps, and scissors. We have not yet learned to build structures with the precision and complexity of these remarkable molecular machines.

V. Just For Fun

Which Has More DNA? An Onion or an Egg?

This experiment is a bit more involved than most you have already done, but you will be taking actual DNA from an onion and an egg! Then you can compare the samples to see which has more DNA. Each step in this experiment has an explanation about what is being done to extract nucleic acids from living tissues. A page is provided at the end of the experiment for you to take notes as you are performing it.

Hints for performing this experiment: Try to make your measurements accurate. Try not to lose too much of the samples as you transfer them from container to container.

Materials

one raw egg
one raw onion
table salt
clear liquid dish washing detergent
rubbing alcohol (isopropanol, 70-90%)
wooden stir stick or Q-tip
coffee filter (any color)
sieve
2 glass jars or large test tubes
measuring cup and measuring spoons
blender

❶ Lysing the cells in the sample

Recall that living tissues are made of cells and the nucleic acids are inside the cells. In order to extract nucleic acids, the cells need to be opened, or *lysed*. The first step of this experiment opens, or lyses, the cells. This is accomplished by using a combination of detergents and enzymes.

Part A: Prepare the sample for lysis

1) Put a shelled raw egg in a blender and add 240 ml (1 cup) cold water. To this mixture, add 5 ml (1 teaspoon) of table salt and blend the sample on high speed until it is pureed (about 15-20 seconds).

2) Pour the egg sample through a sieve into a glass jar or large test tube.

3) Repeat for the onion but this time add .5 liter (2 cups) of water.

4) You need to have at least 60 ml (1/4 cup) of liquid for each sample before continuing with the experiment.

Part B: Lysis

1) Add detergent to the mixtures to break the cells open. For each sample, use 15 ml (1 tablespoon) of detergent per cup of cell mixture.

2) Gently swirl the cell, water, and detergent mixture by rotating the container or mixing with a stick or spoon, being careful not to create foam.

3) Allow the mixture to sit for 5 minutes, gently swirling intermittently.

❷ **Separating the DNA from the cell material**

Once the cells are lysed, or broken open, you'll have a mixture of DNA, RNA, proteins, and other cell parts. The DNA and RNA need to be separated from the rest of the mixture. This is accomplished with the use of alcohol. The nucleic acids are not soluble in alcohol; therefore, they will precipitate out of the solution.

1) Tilt the jar or test tube and slowly add 60 ml (1/4 cup) of isopropanol per 60 ml (1/4 cup) of the cell/water mixture, carefully pouring it down the inside of the jar or test tube. This must be done slowly without agitating the mixture. The alcohol will float to the top of the jar or test tube, and the DNA will precipitate at the water-alcohol interface (the area where the alcohol and water meet).

2) Perform Step 1) for both samples.

❸ **Pulling out the DNA**

After the DNA is separated from the cell parts, it can be extracted, or pulled out of the solution. This is accomplished with the use of a wooden stick or Q-tip. Although all nucleic acids can be removed from the cells, only DNA survives the procedure. RNA is chewed up, or degraded, by enzymes during the process. DNA is more robust than RNA and is not easily degraded by enzymes.

1) Take the Q-tip or wooden stick and insert it into the alcohol layer. Gently touch the alcohol-water interface, and gently swirl the stick or Q-tip, pulling up slightly. The DNA will collect on the stick or Q-tip and long strands should be visible.

2) Continue spinning and collecting the DNA for a few seconds.

3) Pull the stick out and place the DNA on a coffee filter to dry.

4) Write the results and your conclusions in the spaces provided at the end of the experiment.

Commentary

The protocol, or procedure to be followed, in this experiment describes a simple method for extracting DNA from living tissues.

The steps in this experiment can be adapted to use with a variety of different samples by adjusting the volumes for the sample being prepared. If you are lysing animal cells, add 5 ml (1 teaspoon) of meat tenderizer. The enzymes in meat tenderizer are needed to open animal cells.

Troubleshooting

Troubleshooting is part of doing science. Few experiments work the first time. Many new discoveries are made by scientists when their experiments "fail."

Frequently Asked Questions

Q. If I do not get at least 60 ml (1/4 cup) of liquid from Step ❶, Part A, should I continue?

A. No. If you do not get at least 60 ml (1/4 cup) of material, blend the sample again, adding more water. With smaller volumes, there may not be enough DNA *extracted* to be visible.

Q. What if I do not see an alcohol-water interface?

A. If you do not see an alcohol-water interface, try adding more alcohol, being careful not to agitate the sample. If you still do not see an interface, check the concentration of your alcohol, and make sure it is not less than 70% alcohol. *If your alcohol is 70% or more and you still do not see an interface, add twice the volume of alcohol to the sample.* You should eventually see an interface. If this fails, discard the sample and start over, making sure you use the amount of water specified in Step ❶.

Q. What if I see foam?

A. Carefully remove the foam with an eyedropper *without agitating the sample.*

Q. What if I do not get any DNA?

A. There could be several reasons you do not see any DNA.

1) You did not use enough starting material. Repeat the experiment and double the amount of starting material.

2) You did not use enough detergent. Repeat the experiment using more detergent.

3) You did not use the right kind of detergent. Repeat the experiment with a different detergent.

4) You did not let the sample sit long enough to break open the cells. Repeat the experiment and allow the sample to sit for a longer period of time.

5) You did not add detergent or enzymes to the sample. Repeat the experiment adding enzyme or detergent or both.

6) You did not add enough alcohol. Add more alcohol.

7) The alcohol you added was not concentrated enough. Add 70-90% rubbing alcohol.

8) There is not enough salt in the water mixture to precipitate the sample. Add 5 ml (1 teaspoon) of table salt to your water-cell-alcohol mixture. Swirl. Add more rubbing alcohol until you see an interface, then try to pull out the DNA.

Note: This experiment can be repeated using samples of other living things including but not limited to:

- Vegetable tissue such as spinach, peas, green beans, broccoli, onions, etc.
- Grains such as wheat germ, corn, oatmeal, seeds, or yeast.
- Animal tissue such as eggs, chicken or beef livers, chicken hearts, etc.

All items must be *uncooked*.

Notes

_____

_____

_____

_____

_____

_____

_____

_____

_____

_____

_____

_____

_____

_____

_____

_____

CHEMISTRY

CHEMISTRY

Results

Conclusions

What conclusions can you draw from your observations?

_____

_____

_____

_____

_____

_____

_____

_____

_____

_____

_____

_____

_____

_____

_____

_____

Experiment 6

Identifying Plants

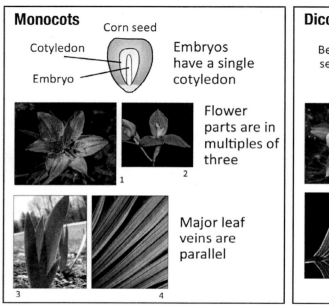

Monocots

Corn seed

Cotyledon

Embryo

Embryos have a single cotyledon

Flower parts are in multiples of three

Major leaf veins are parallel

1

2

3

4

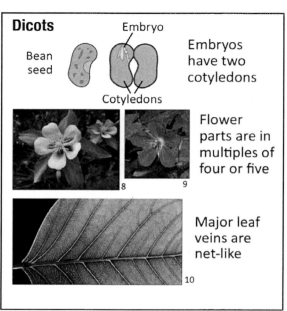

Dicots

Embryo

Bean seed

Cotyledons

Embryos have two cotyledons

Flower parts are in multiples of four or five

Major leaf veins are net-like

8

9

10

Introduction

Explore identifying plants by observing them in their ecosystem.

I. Think About It

❶   What types of plants do you see growing in your neighborhood?

_____

_____

_____

_____

❷   How would you describe the climate you live in?

_____

_____

_____

_____

❸   Do you think you could find moss in your neighborhood? Why or why not?

_____

_____

_____

_____

❹ Do you think you could find conifers in your neighborhood? Why or why not?

_____

_____

_____

_____

❺ Why do you think you can find some types of plants in your neighborhood but not others?

_____

_____

_____

_____

❻ How do you think you could tell if a plant you are observing is a vascular or non-vascular plant?

_____

_____

_____

_____

_____

_____

II. Experiment 6: Identifying Plants   Date _____

Objective _____

_____

_____

Materials

colored pencils
handheld magnifying glass
field notebook (begin a field notebook or use an existing one)
backpack with water and snacks

Optional

field guide to the plants book
iPad, camera, or smartphone with camera
plant identification app

EXPERIMENT

❶ Pick a location where you can observe different kinds of plants. If you live in the city, you might arrange a trip to a local park or a hike somewhere outside the city limits.

❷ Before you go, prepare your field notebook by writing down the plant classification information from the chart in Section 6.3 of the *Student Textbook*. You can use the common names for the phyla. List the features of monocot and dicot leaves and flowers from the chart in Section 6.7. Tear out the Plant Identification Guide at the end of the *Results* section of this book and stick it in your field notebook.

❸ Spend a few hours walking or hiking in the location you've chosen. As you walk, observe the plants around you.

❹ Find a small area where different kinds of plants are growing. Observe the plants.

❺ Choose 3-5 plants and observe them more closely. These can be any type of plant: grasses, mosses, flowering plants, trees, etc.

❻   In your field notebook, draw or sketch the plants you have chosen. Drawing plants helps you observe their features more closely. Note details such as the shape of the leaves, how they are arranged, and what pattern the veins make. Draw or sketch the whole plant first. Then select a small area to examine with a magnifying glass and draw details. Next to the drawing, or as labels on the drawing, make written notes about some of the features you observe. In addition, you can also photograph the plants and later print the photos and attach them in your field notebook. Also make notes about features of the type of ecosystem the plant lives in. For example, shady, sunny, moist, dry, sandy soil, etc.

❼   To help identify each plant by its type, refer to the notes you made about plant classifications and monocot vs. dicot features and use the Plant Identification Guide. You can also use a field guide to the plants book or use a plant identification app. Record your identifications in your field notebook.

Results

See field notebook.

Plant Identification Guide

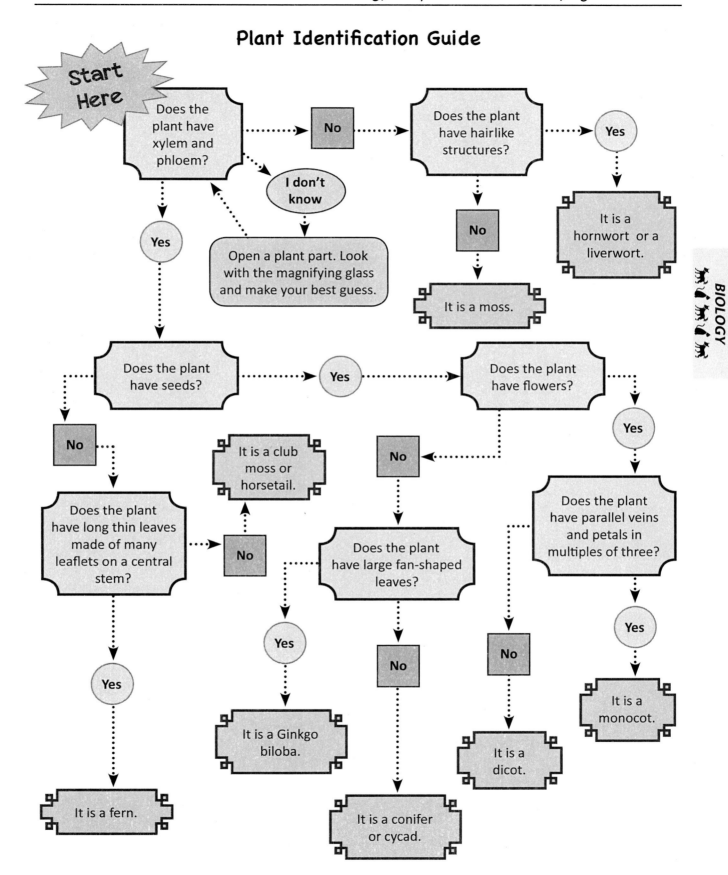

BIOLOGY

III. Conclusions

What conclusions about plant identification can you draw from your observations and the data you collected? What did you discover about plants and their ecosystems?

_____

_____

_____

_____

_____

_____

_____

_____

_____

_____

_____

_____

_____

_____

_____

_____

_____

BIOLOGY

IV. Why?

While doing this experiment, you probably found that you usually pass plants by without really noticing more than the most obvious features—what color they are, how tall or short, whether they are grassy or leafy. By stopping and looking closely, you can begin to observe details such as the shape of the leaves and how they are arranged. Looking at plants closely helps us appreciate how varied they are and how their structures differ to allow them to live in particular ecosystems.

If we observe certain types of plants growing together and we know what conditions they need to grow, we can draw some conclusions about the amount of water, the weather, and the type of soil that most likely exist in that area. And knowing something about the amount of water, the weather, and the type of soil in an ecosystem can tell us what kinds of plants we are likely to find there. If you are a gardener or a landscaper, this knowledge is important in order for you to be able to choose plants that will grow well in a particular location. What conditions does a particular plant need to grow? Will it get the proper amount of water? Will it get enough sun or shade? Will it survive winter cold or summer heat? Will it have the proper soil to meet its needs?

Carefully observing plants over time leads to many discoveries in many areas of science. One observation can lead to more questions, such as what does this plant look like as a seedling, a young plant, and a mature plant that has reproductive structures? How does it change throughout the seasons? Does the plant look healthy or is it lacking something in its environment, such as sufficient water or sunlight? Which insects are attracted to a particular plant to get nectar and to pollinate the plant in return? Which insects eat a plant and at what part of the insect's life cycle? How does wildlife interact with the plants in an ecosystem? Which animals eat which plants? How do animals make their homes in and among plants? Do the animals live in the ecosystem year-round or do they migrate? Are human activities affecting the ecosystem and the plants and animals that live there?

You'll also begin to see how everything fits together in an ecosystem, with plants providing the base for other living things. Without plants, other life forms on Earth could not exist.

V. Just For Fun

Monocot vs. Dicot

Recall that corn is a monocot and beans are dicots. In this experiment you'll have a chance to observe differences between these two types of plants.

Materials

> 2 plant pots
> potting soil
> corn seeds, 8 or more with packet
> bean seeds, 8 or more with packet
> field notebook
> colored pencils
> magnifying glass
> water
> warm, sunny location

❶ Get two plant pots and fill them with potting soil.

❷ Look at the seed packets for the corn and the beans to find the estimated number of days the seeds will take to germinate. Based on the germination rate shown on the seed packets, try to figure out how many days apart you will need to plant the corn and bean seeds so they are likely to germinate on the same day. Generally, beans take a little longer to sprout than corn. Record your estimate in your field notebook.

❸ Plant 8 or more corn seeds in one pot and 8 or more bean seeds in the other pot according to your estimate in Step ❷. Label the pots. Put the pots in a warm, sunny location and water thoroughly. Check the soil daily, add water as needed to keep the soil moist, and look for sprouts.

❹ Note when the sprouts begin to break the soil. Record your observations in your field notebook. Include the dates when sprouts break the ground and how many of each type. Did corn and bean sprouts break the soil on the same day? What do they look like? Can you observe any differences? Can you see more details with a magnifying glass? Continue to observe the plants daily.

BIOLOGY

❺   Pull up one corn sprout and one bean sprout that have broken ground and observe the roots, stems, and leaves. Use a magnifying glass to see more details. Every few days pull up another sprout of each type to check the root development until you have 1 or 2 plants left that you can let grow. Record your observations in your field notebook for each plant you pull up.

❻   Observe the sprouts as they grow taller and develop more leaves. Record your observations, noting how the features of each type of plant change over time. Also note the similarities and differences between the two types of plants.

❼   When you have 1 to 2 plants left that are still in the soil, continue to tend them and observe how they grow and change. Record your observations.

Experiment 7

Take Away the Light

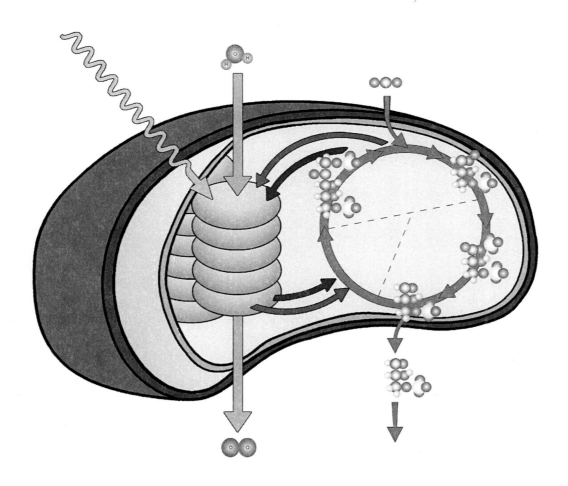

Introduction

Explore photosynthesis!

I. Think About It

❶　Where do you think plants get their food and nutrients? How do they get them?

_____

_____

_____

_____

_____

❷　Do you think some types of plants need more sunlight than other types of plants? Why or why not?

_____

_____

_____

_____

❸　Why do you think different types of plants have different kinds of leaves?

_____

_____

_____

_____

BIOLOGY

❹ What do you think Earth would be like if there were no plants? Why?

_____

_____

_____

_____

_____

❺ What do you think would happen to a plant if it didn't get water for a long time? Why?

_____

_____

_____

_____

_____

❻ What do you think would happen to a plant if bugs ate all its leaves? Why?

_____

_____

_____

_____

_____

BIOLOGY

II. Experiment 7: Take Away the Light   Date _____

Objective   _____

_____

Hypothesis   _____

_____

_____

BIOLOGY

Materials

plant with at least 6 flat, green leaves
lightweight cardboard or construction paper
scissors
tape
2 small jars
marking pen

EXPERIMENT

❶   Take some cardboard or construction paper and cut it into pieces large enough to completely cover a leaf. Make enough pieces to cover the front and back of 3 leaves.

❷   Six different leaves will be tested. Two of the leaves will be left on the plant (attached) and four leaves will be removed from the plant (unattached).

❸   Using the cut out cardboard pieces from Step ❶, tape two pieces so they cover the front and back of one of the leaves attached to the plant. Another leaf will stay attached to the plant and remain uncovered.

❹   Remove four leaves from the plant. Cover two of them with cardboard. These unattached leaves will be either: covered with cardboard and the stem placed in water, covered out of water, uncovered with the stem in water, or uncovered out of water.

❺ With the marking pen, label the leaves in the following manner:

Leaf 1:   **UA** — uncovered, attached

Leaf 2:   **CA** — covered, attached

Leaf 3:   **UUW** — uncovered, unattached, in water

Leaf 4:   **CUW** — covered, unattached, in water

Leaf 5:   **UU** — uncovered, unattached (no water)

Leaf 6:   **CU** — covered, unattached (no water)

❻ Take two small jars and fill them with water. Take the two leaves that will be placed in water and prop one in each jar, keeping the stem submerged and the leaf out of the water. Check the water level every day over the course of the experiment to make sure there is enough water in the jars for the stems to be submerged.

❼ Wait several days, and then every day observe the changes to the leaves. Carefully remove the cardboard from the covered leaves to make your observations, and then re-tape the cardboard. Record your observations in the chart in the *Results* section.

BIOLOGY

BIOLOGY

Results	UA	CA	UUW	CUW	UU	CU
Day 1						
Day 2						
Day 3						
Day 4						
Day 5						
Day 6						
Day 7						
Day 8						
Day 9						
Day 10						

III. Conclusions

What conclusions can you draw from your observations?

_____

_____

_____

_____

_____

_____

_____

_____

_____

_____

_____

_____

_____

_____

_____

_____

_____

IV. Why?

Because plants don't have teeth, a mouth, a stomach, or saliva, they can't eat food the way animals do. Instead, plants are photosynthetic and use the Sun's energy to make their own food. However, there are a few carnivorous and parasitic plants that obtain additional nutrients by non-photosynthetic means.

Photosynthesis occurs in specialized organelles called chloroplasts that contain chlorophyll molecules found in the thylakoid, a membrane-bound compartment in the chloroplast. The chlorophyll molecules "pick up" the light energy and transfer this energy to other molecules in a series of reactions that ultimately produce sugar molecules. This is an example of "light energy" being converted into "chemical energy."

It is hard to overemphasize the importance of photosynthesis to all living things. All creatures need a source of energy or they will die. Plants use photosynthesis to make their food from energy from the Sun. We, and all other animals, get our energy by eating plants or by eating other animals that eat plants. The Sun's energy is ultimately the source of energy for all living things. Plants put this energy into a form usable by animals, including ourselves.

In this experiment you explored how plants need light to grow and stay healthy. Leaves are the main organ for carrying out photosynthesis in plants. Two of the leaves you used in this experiment are "controls." The leaf that is attached and uncovered is a positive control (Leaf 1). This leaf should remain healthy throughout the course of the experiment unless something happens to the plant. The leaf that is covered and unattached and without water is a negative control (Leaf 6). This leaf should die first.

Positive controls help the investigator determine if the experimental setup is working. Negative controls tell the investigator when the desired effects of the experiment have indeed occurred. Both types of controls are useful for making sure the experimental results are valid. In other words, if the positive control did not work (the leaf died on the plant), then when the other leaves die, it could not be concluded that the leaves died because of the changes made during the experiment. Likewise, the negative control should be a negative result (the leaf should die without water or sunlight) to prove that the other samples survive or die accordingly. If the negative control survives, it is an indication that something else is happening. For example, the leaves are especially tough, the experiment needs more time, a plastic plant was used instead of a real one and so the leaves cannot die (anything is possible). Positive and negative controls give the experimenter boundaries for the experiment and allow valid conclusions to be made about the samples in between.

BIOLOGY

V. Just For Fun

❶ Get 4 or more plant pots and fill them with potting soil. In each pot, plant 3-4 bean seeds and add enough water to thoroughly moisten the soil.

❷ Label each pot for reference when recording your observations.

❸ Put the pots in a warm, sunny location and keep the soil moist until the beans begin to grow into seedlings.

❹ Place the pots with the seedlings in different locations with varying amounts of sunlight—from no sunlight to full sun. The pots can all be placed inside or all outside. Water them regularly.

❺ Think about what information you will need to record during the experiment. In the following space or in your field notebook or on separate paper, make up your own chart for recording your observations. Include the plant identifier (number) and how much sunlight the plant will get in its particular location.

Observe the plants for several weeks and compare their growth and health. Record observations for all the plants every few days.

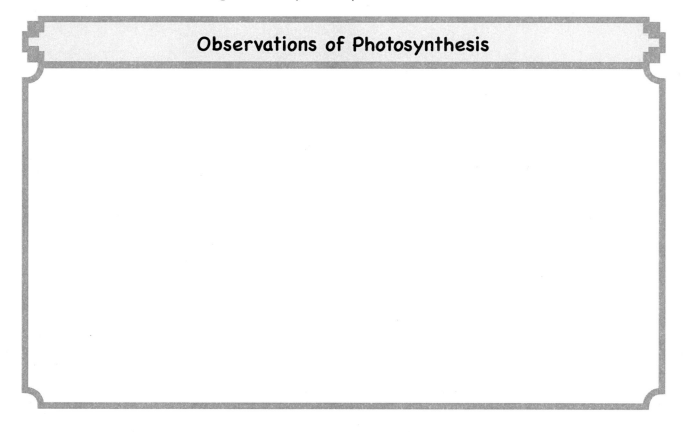

Observations of Photosynthesis

Observations of Photosynthesis

Observations of Photosynthesis

Conclusions

What conclusions can you draw from your observations?

_____

_____

_____

_____

_____

_____

_____

_____

_____

_____

_____

_____

_____

_____

BIOLOGY

Experiment 8

Seeing Inside Plants

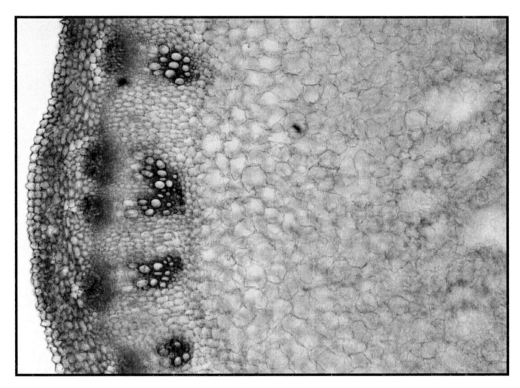

Cells in a stem

Introduction

Use your microscope to see what you can discover about plant cells and tissues.

I. Think About It

❶ Do you think a plant could live and grow if all its cells were the same? Why or why not?

_____

_____

_____

_____

_____

❷ Why do you think plants have stems?

_____

_____

_____

_____

_____

❸ What do you think a plant uses its leaves for?

_____

_____

_____

_____

_____

BIOLOGY

❹ How do you think plants get the nutrients they need?

_____

_____

_____

_____

_____

❺ What do you think you would discover if you could see how water travels through a plant?

_____

_____

_____

_____

_____

❻ What do you think would happen to a tree if a small area of its xylem layer were trimmed off? An area that went all the way around the tree? Why?

_____

_____

_____

_____

_____

II. Experiment 8: Seeing Inside Plants   Date _____

Objective   _____

_____

Hypothesis _____

_____

_____

Materials

microscope with 4X, 10X, and 40X objective lenses; a 100X objective lens is
   recommended
glass microscope slides (plain)
glass microscope coverslips
immersion oil (if using 100X objective lens)
water
eyedropper
sharp knife
toothpick
colored pencils
Samples:
   raw celery stalk with leaves
   raw carrot
   a large leaf
   other plant parts from plants of your choice

Preparing Wet Mount Slides

The samples in this experiment should be prepared with wet slides, which are also called
wet mounts. The water in a wet mount slide helps support the sample and can also help
flatten it and make it more translucent. In addition, by filling the space between the
coverslip and the slide, the water allows light to pass more easily through the sample.

A wet mount is made by simply putting a drop of water in the middle of a plain slide and
gently placing the sample on top of the water. A coverslip is then placed over the sample by
placing one edge of the coverslip at the edge of the water on the slide and carefully lowering
the coverslip over the sample. A toothpick can be used for lowering the coverslip. Gently

lowering the coverslip in this way will prevent air bubbles from forming under it. Having some air bubbles is okay, but too many will make it hard to see the sample.

If there is too much water on the slide and the coverslip won't stay in one place, you can hold a piece of paper towel at the edge of the slide and coverslip to draw out some of the water. If there is too little water under the coverslip, you can put a drop of water right next to the coverslip and some of the water will run underneath it. It takes a little practice to know how much water to use.

EXPERIMENT

Note: If it's been a while since you used your microscope, check out the refresher instructions at the end of this chapter.

❶ Take a stalk of celery. Have an adult help you cut a very thin cross sectional slice of the stem. Make the slice as thin as possible so light can pass through it.

❷ Make a wet mount of your celery sample.

❸ Place the sample in the microscope and look at it with different magnifications starting with the lowest. What different tissues and cells can you see? You can refer to the *Student Textbook* to help with the identification.

Remember that when you are using a 100X oil immersion lens, you need to turn the turret until the oil immersion lens is halfway into position. Place a single drop of immersion oil on the glass coverslip and gently move the oil immersion lens in place.

[**NOTE: It is extremely important that the lens does NOT scrape the coverslip. This can scratch the lens and ruin it.** The lens will be very close to the glass coverslip if you have adjusted the focus correctly with the other lenses. If it seems like the lens will scrape the coverslip, gently back up the lens with a few turns of the fine adjust knob.]

❹ In the *Results* section draw what you see. Label what you think the different structures are and what kind of cells they are made of.

Based on your observations, is celery a monocot or a dicot? Why?

_____

_____

❺ Repeat Steps **❶**-**❹** with a carrot.

❻ Compare what you observed in the samples of the celery and the carrot. What part of the plant is each sample? How are they the same and how are they different?

❼ Take two celery leaves. Make a wet mount slide of one leaf and a dry mount slide of the other. Which type of slide works better for this sample? Why?

Repeat Steps **❸** and **❹**. Can you observe epidermal cells? Stomata? What else can you observe about the outside of a leaf that you cannot see with just your eyes?

❽ Take a leaf from a plant that has large leaves. Have an adult help you prepare the thin section. Cut a strip about 1" wide across the middle of the leaf. Roll the strip up and cut a few very thin slices from the roll. Pick the best slice and make a wet mount sample so you are looking at the cut edge of the leaf with the midvein of the leaf at the center of the slide. Record your observations.

Step **❽**

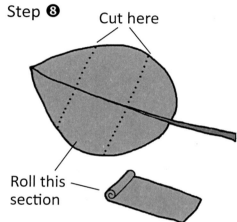

Cut here

Roll this section

❾ Using the appropriate experimental steps above, make slides of other plant tissues and record what you see. In addition to stems, roots, and leaves, you can make samples of parts of flowers, fruit, and seed pods. Record your observations in this book, in your field notebook, or on separate sheets of paper.

Results

Observations of a Celery Stalk

Observations of a Carrot

Celery and Carrot Comparisons

BIOLOGY

Observations of the Underside of a Celery Leaf

Dry mount or wet mount? _____

Observations of the Interior of a Leaf

BIOLOGY

Observations of Other Plant Samples

III. Conclusions

What conclusions can you draw from your observations?

_____

_____

_____

_____

_____

_____

_____

_____

_____

_____

_____

_____

_____

_____

_____

_____

IV. Why?

A microscope makes it possible to observe the cells and tissues that plants are made of, which helps us begin to understand what complicated organisms plants are.

The stalk is the stem of the celery plant. You were probably able to observe the main tissues of the celery stem: xylem, phloem, and pith. The xylem and phloem tissues are responsible for transporting food, minerals, water, and other nutrients from the roots to the leaves and from the leaves to the roots. In a dicot such as celery, the xylem and phloem are arranged in vascular bundles localized near the epidermal tissues and arranged in a ring-like pattern rather then being scattered throughout the ground tissue of the stem.

The xylem transports water and minerals from the roots to the leaves at a rate of about 15 meters (49 feet) per hour for most plants. The fluid in the xylem rises against gravity without the use of molecular motors or pumps by means of evaporation of water from the leaves through stoma. As water evaporates from the leaves, a negative tension, or pressure, is created that pulls the water upward from the roots. This is called transpirational pull.

The transport of sugars in the phloem works by a different mechanism than the transport of minerals in the xylem. The phloem sap moves downward through a series of sieve tubes connected to each other end to end. The sugar made by photosynthesis in the leaves is transported with small protein pumps that pump it into the phloem tissue.

The xylem and phloem in a dicot stem are arranged in a ring surrounding core tissue called the pith which is made mostly of parenchyma cells. The xylem and phloem in monocots is arranged in vascular bundles that are scattered throughout the ground tissue.

The part of a carrot that we eat is the taproot of the plant. A carrot is a dicot, and you may have been able to observe the xylem tissue in the core of the carrot and the phloem tissue surrounding it, which is typical of a dicot root. The lines extending from the core of the carrot outward are called pith rays and are made primarily of parenchyma cells. In a monocot root the xylem and phloem are found together in the center of the root.

In the cross section of the leaf, you may have been able to see xylem and phloem in the leaf vein, mesophyll tissues making up the major portion of the leaf, and epidermal tissue on the outside.

BIOLOGY

V. Just For Fun — Colorful Flowers   Date _____

Objective   _____

_____

Hypothesis   _____

_____

Materials

3 or more small jars
several fresh white carnation flowers
food coloring
sharp knife
colored pencils
microscope

EXPERIMENT

❶   Take two or more of the small jars and add water and several drops of food coloring to each. Use a different color in each jar.

❷   Trim the end of one carnation stem at an angle, and place it in one of the jars of colored water.

❸   Check the petals of the carnation every couple of hours, and in the *Results* section record any color changes observed.

❹   Take the carnation out of the jar, and cut a thin slice of the stem. Make a wet mount and look at the sample under the microscope. Can you identify the xylem and the phloem? What other structures and cells can you observe? On the second page of the *Results* section, draw what you see and add labels to the drawing.

❺   Take one stem and slice it lengthwise with a knife, starting about halfway up the stem and cutting away from the flower. (Have an adult help you.) Stick one end of the divided stem into a jar of colored water and place the other part of the stem in a jar that contains water of a different color. Let the carnation soak up the colored water until the petals begin to change color. In the *Results* section, draw what you observe.

Results

BIOLOGY

Color change in carnation placed in one color water

Color change in split carnation placed in two colors of water

Xylem, phloem, and other stuff through the microscope

III. Conclusions

What conclusions can you draw from your observations?

_____

_____

_____

_____

_____

_____

_____

_____

_____

_____

_____

_____

_____

_____

BIOLOGY

Refresher—How to Use a Microscope

❶   Move the microscope to a desk or table where you can sit for a few hours.

❷   Turn the revolving turret so the lowest power objective lens (4X) is clicked into position. [**NOTE: Be extremely careful not to bang the lenses on the stage as you turn the turret. This can damage the lenses.**]

❸   Turn the coarse adjust knob and observe how the stage moves up and down.

❹   Turn the fine adjust knob and observe how the stage moves up and down.

❺   Turn on the light source and use the condenser to change the amount of light entering the stage.

Observing a Sample

1.   Put your sample in the microscope.

2.   With the lowest power objective (4X) in place, look through the ocular lens, and using the coarse adjust knob, slowly move the objective lens up and down until the sample is in focus.

3.   If necessary, move the slide to bring your sample into view.

4.   Turn the fine adjust knob slowly up and down as you observe the sample. You should see some parts come into focus as other parts go out of focus. Notice the range of focus (how much of the sample is in focus).

5.   Keeping the microscope steady, gently turn the turret until the next highest power objective lens (10X) clicks into place. Be careful not to bump the objective lens into the sample. If this looks like it will occur, move the lens up with the coarse adjust knob.

6.   Look through the ocular lens and observe the sample, turning the coarse and then fine adjust knobs until the sample comes into focus. Notice the range of focus.

7.   Repeat Steps 4 and 5 with the 40X lens.

8.   If you have a 100X oil immersion lens, turn the turret until the oil immersion lens is halfway into position. Place a single drop of immersion oil on the glass cover slip and gently move the oil immersion lens in place.

[**NOTE: It is extremely important that the lens does NOT scrape the coverslip. This can scratch the lens and ruin it.** The lens will be very close to the glass coverslip if you have adjusted the focus correctly with the other lenses. If it seems like the lens will scrape the cover slip, gently back up the lens with a few turns of the fine adjust knob.]

9.   Adjust the condenser if more light is needed, and using the fine adjust knob only, focus the image. [**Never use the coarse adjust knob to focus an oil immersion lens — it is too easy to smash the lens against the slide.**]

10. Rotate the turret to move the oil immersion lens away from the sample. With the coarse adjust knob, lower the stage and remove the sample.

BIOLOGY

Experiment 9

Growing Vegetables from Scraps

Courtesy of National Institutes of Health

Introduction

Find out if vegetables can use vegetative reproduction to grow into new plants.

I. Think About It

❶ How many different types of vegetables do you have in your refrigerator or sitting on your kitchen counter? What are they? What vegetables do you wish you had?

_____

_____

_____

_____

_____

❷ Where did these vegetables come from? How did they get to your kitchen?

_____

_____

_____

_____

❸ Do you think you could regrow some of these vegetables? If so, how would you do it?

_____

_____

_____

_____

❹ Do you think if you could regrow a vegetable it would be edible? Why or why not?

_____

_____

_____

_____

_____

❺ Do you think vegetable gardeners use vegetative reproduction methods to grow plants? If so, why do you think they do?

_____

_____

_____

_____

_____

❻ What advantage do you think plants have if they can reproduce sexually with seeds and can also reproduce using vegetative reproduction?

_____

_____

_____

_____

_____

BIOLOGY

II. Experiment 9: Growing Vegetables from Scraps Date _____

Objective _____

_____

Hypothesis _____

_____

_____

Materials

select several vegetable scraps such as: carrot top, lettuce leaves or root end of a head
   of lettuce, red beet top or root, turnip top or root, garlic bulb, onion bulb, scallions,
   either or both ends of a zucchini squash or cucumber, basil leaves with stem, potato
   (piece or peeling), other vegetables of your choice
   Note: If there are leaves on a vegetable top, such as a carrot or red beet, these should
   be cut off.
knife
toothpicks
several small glass jars or small drinking glasses
colored pencils or pens

Optional

field notebook

EXPERIMENT

❶ Select several vegetable scraps that are typically discarded during meal prep. See
   suggestions in the *Materials* section. Experiment with different types of cuttings. Leave
   enough of the vegetable that you can put toothpicks in it—about 1.5-3 cm (.5-1 inch).
   Use fresh cuttings for best results.

❷ Gather enough small glass jars or small drinking glasses to hold the vegetable cuttings.

❸ Decide which part of each cutting will be placed downward in a jar of water. If you are
   unsure, you can experiment by using two cuttings with each in a different orientation.
   The cuttings will need to be suspended from the top of a jar so they are in water but not

submerged. You can do this by sticking toothpicks into the sides of the cutting and then putting your sample in the jar with the toothpicks resting on the jar rim.

❹ Place the cuttings in an area where they will stay warm and get plenty of sun and then fill the jars with enough water to cover the bottom and part way up the sides of each cutting.

❺ Check the cuttings daily to make sure the bottoms are still in water. Change the water occasionally if needed.

❻ Over the next several weeks observe how the cuttings grow. Do they grow leaves and roots? Only roots? Only leaves? Do they become a full vegetable? Do they grow at all? Record your observations in the *Results* section or in your field notebook.

Results

Record your results in the following boxes or in your field notebook. Make a section for each plant you are experimenting with. Record the type of plant, the part of the plant you think the cutting contains, and the date you put it in water. Record your prediction of whether or not you think a plant will grow from the cutting and why. Check your samples daily to make sure they have enough water. When you notice a change in a cutting, record your observations in writing and drawings and note the date. If you run out of space, you can fasten more paper in this section.

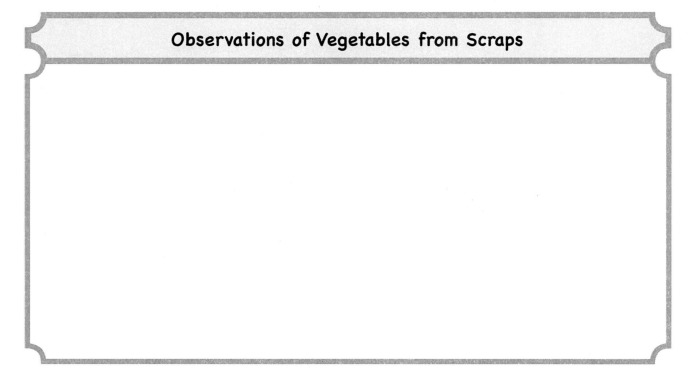

Observations of Vegetables from Scraps

More Observations

BIOLOGY

Even More Observations

BIOLOGY

III. Conclusions

What conclusions can you draw from your observations? How easy or difficult was it to grow vegetable plants from cuttings? What problems did you run into? Did some cuttings grow and some not? What do you think you might do differently?

_____

_____

_____

_____

_____

_____

_____

_____

_____

_____

_____

_____

_____

_____

BIOLOGY

IV. Why?

Plants are amazing organisms. In order to survive being eaten and stepped on and going through droughts and fires, plants have had to develop varied features. Some have buds on roots or rhizomes that can grow into new plants even when they are detached from the original plant. Some have bulbs or corms that help the plant survive drought and allow the plant to regrow in the spring after the stems and leaves have died over the winter. Most will grow leaves when some of the existing leaves have been eaten or cut off.

There are plants that need fire in order to survive. Tallgrasses that grow on prairies have roots that grow deep enough into the ground to survive the heat of fire at the surface, and new plants will grow from the roots after the fire has passed. The fire clears out the old dead plant parts leaving space for new plants to grow and releasing nutrients that were stored in the dead matter. Giant sequoia trees need fire to survive because their cones won't open to release seeds without being exposed to the intense heat. Without fire there can be no sequoia seedlings. The trees themselves have very thick bark, as much as two feet thick near the base, that allows them to go through fires without damage.

Still other plants are adapted to survive drought. Prickly pear cacti have flat thick pads that are filled with fluid when there is moisture, but when it is very dry, the pads wither as the plant draws on the stored fluids. The pads can get so withered that they begin to look dead, but if moisture arrives soon enough, the roots will bring water and nutrients to the pads, plumping them up and restoring them to health.

In this experiment you explored how some different plants use vegetative propagation by growing roots, stems, and leaves from a small cutting of the plant. You probably found that cuttings of some plants were able to grow into an entire new plant, while others may have grown only parts of a plant or may have not been able to grow at all and died.

Plants have different structures that enable them to grow from a cutting. For example, onions and garlic bulbs have a basal plate from which roots can grow, and they contain a bud that can grow into the shoot system. A new onion or garlic plant can grow from a bulb or the lower part of a bulb. From the top section of a carrot, the plant can grow new stems and leaves and new side roots, but once the taproot is severed from the plant, it has no meristematic tissue from which a new taproot can grow. By learning about the different structures of plants, you will be better able to determine whether you can use vegetative propagation to grow new plants from plants already existing in your outdoor or container garden.

V. Just For Fun

Eating Scraps

Once each cutting has grown enough roots, plant it in a pot filled with potting soil. Check the soil frequently to make sure it stays moist but not too wet and keep the plant in a warm, sunny location indoors or outdoors.

Record the growth and health of your plants and the dates of your observations. Do the plants get big enough for you to harvest to eat? How long does it take? Which ones grow best? Which ones die or don't grow well? Why do you think you got these results?

Note: Potato stems, leaves, and roots contain a toxin called solanine and should not be eaten. Potatoes that are green under the skin should also not be eaten, but potatoes that are not green are perfectly safe to eat.

Another note: Carrot leaves are edible and recipes can be found on the internet. Or feed them to a nearby horse!

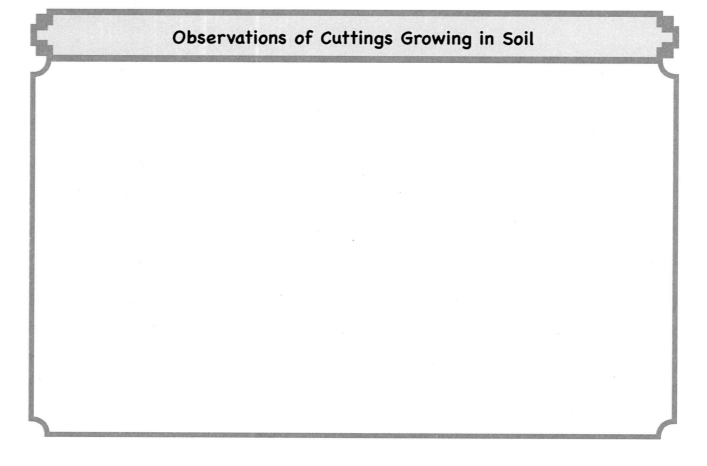

Observations of Cuttings Growing in Soil

More Observations of Cuttings Growing in Soil

BIOLOGY

More Observations of Cuttings Growing in Soil

Experiment 10

Power Pennies

Voltaic Cell

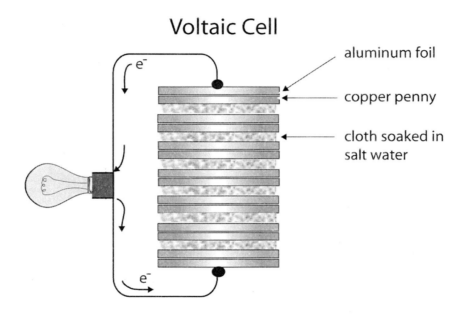

aluminum foil

copper penny

cloth soaked in
salt water

Introduction

Explore chemical energy by constructing a voltaic battery.

I. Think About It

❶   How do you think a voltaic battery works?

_____

_____

_____

_____

_____

❷   What do you think would happen if you put the parts of a layer of a voltaic battery in a different order? Why?

_____

_____

_____

_____

_____

❸   How can you use the chemical potential energy in a battery? Why?

_____

_____

_____

_____

❹ How do you think the invention of the battery changed the way people live?

_____

_____

_____

_____

_____

❺ How many things can you think of that are powered by batteries?

_____

_____

_____

_____

_____

❻ What do you think the difference is between mechanical energy and chemical potential energy?

_____

_____

_____

_____

_____

PHYSICS

II. Experiment 10: Power Pennies

Date _____

Objective _____

_____

Hypothesis _____

_____

_____

Materials

10-20 copper pennies
aluminum foil
paper towels
salt water: 30-45 ml (2-3 Tbsp.) salt per 240 ml (1 cup) water
voltmeter
2 plastic-coated copper wires, each 10-15 cm (4"-6") long
duct tape (or other strong tape)
scissors
wire cutters
fine steel wool

Optional

wire stripping tool

EXPERIMENT

❶   Scrub the pennies with steel wool.

❷   Cut out up to 20 penny-size circles from the aluminum foil and from a paper towel. It is important that the cutouts be very close to the size of a penny.

❸   Soak the paper towel circles in the salt water.

❹   Strip the plastic coating off both ends of one piece of wire. Use wire cutters to carefully cut through the plastic without cutting the metal wire or use a wire stripper. Use a small piece of tape to fasten one end of the wire with exposed metal touching the penny.

❺ Strip the plastic from the ends of the second piece of wire. Tape the exposed metal on one end of the wire to a piece of aluminum foil.

❻ Begin stacking the pieces. Place the circle of aluminum foil on a firm surface with the attached wire touching the surface. Put one of the wet paper towel circles on top of the aluminum foil. The paper towel piece should be wet but not dripping. On top of the paper towel, place the penny that has the wire taped to it. It should look like this:

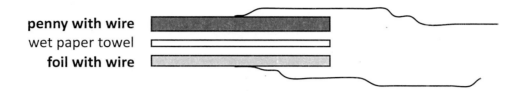

❼ Take the battery wires and connect them to the leads (wires) of the voltmeter. Switch the voltmeter to "voltage" and in the *Results* section record the number it shows. This is the amount of voltage the single layer battery produces.

❽ Add another "cell" to the battery in between the penny with the wire and the foil with the wire, which are the "ends" of the battery. A cell is a penny layer, a paper layer, and a foil layer.

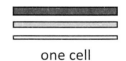

one cell

The battery now has two cells. It should look like the following:

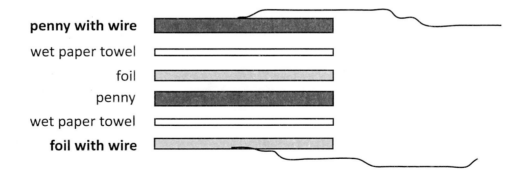

Check the voltage and record the amount in the *Results* section.

❾ Continue adding cells made of foil, wet paper towel, and pennies. Check and record the voltage after each new cell is added. Add as many cells as possible. If your results seem inconsistent, check the construction of your battery.

Results

Number of Cells	Voltage
1	
2	
3	
4	
5	

PHYSICS

Plot your data. Make a graph with "Voltage" on the x-axis and "Number of Cells" on the y-axis.

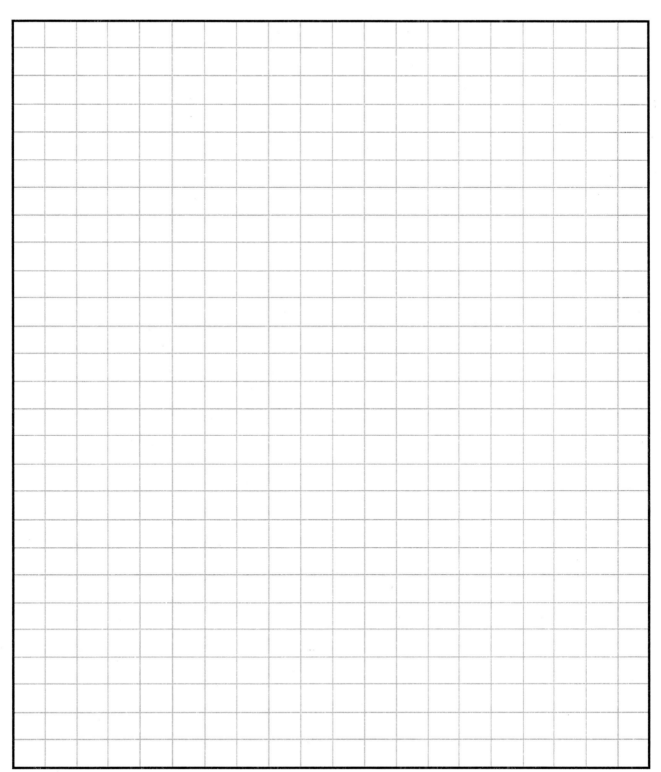

Discuss your data. What can you observe about the data? Discuss any possible sources of error. Evaluate what worked and what didn't work.

_____

_____

_____

_____

_____

_____

_____

_____

_____

_____

_____

_____

_____

_____

PHYSICS

III. Conclusions

What conclusions can you draw from your observations?

_____

_____

_____

_____

_____

_____

_____

_____

_____

_____

_____

_____

_____

_____

PHYSICS

IV. Why?

In this experiment you made a battery based on Alessandro Volta's discoveries. It is thought that Volta was one of the first to observe the principles of electrochemistry, which is the scientific discipline that studies how electricity produces chemical changes and how chemical changes produce electricity.

In the late 1700s Luigi Galvani, an Italian physician, found that the muscle in a dissected frog's leg would twitch when the scalpel he was using touched a nerve while the leg was being held by a copper hook. Galvani felt that he had discovered the effects of "animal electricity," or electrical force stored in the tissues of animals. The Italian scientist Alessandro Volta reviewed Galvani's experiments and thought the result might actually be caused by the contact of the two metals with the moist frog tissues. Volta's experimentation led to the invention of the first electric battery, sometimes called the voltaic pile. Similar to the battery you constructed, Volta's battery was made of alternating discs of leather or paper soaked in a salt solution (or vinegar) and layered between copper and zinc discs. Volta found that an electric current could be generated in this way.

A battery is specifically designed to convert chemical potential energy into electrical energy by means of chemical reactions. In each cell of your battery, the aluminum foil has a chemical reaction with the salt water which causes the aluminum to give up electrons, thus becoming positively charged. The copper penny also has a chemical reaction with the salt water and takes on electrons, becoming negatively charged.

Salt water is a conductor and electrons can be conducted through the salt water from the aluminum foil to the copper penny. In the adjoining cell, the copper penny is touching the foil . Electrons move from the negatively charged copper to the positively charged aluminum which then reacts with the salt water and the process continues through all the cells. When you attach a wire to the aluminum foil at one end of the battery and a second wire to the penny at the other end of the battery and then attach the ends of these wires to a voltage meter, you have made an electrical circuit that allows the electrons to flow from the aluminum foil at one end of the battery through all the cells to the penny at the other end. The electrons then flow from the penny through the wire to the voltage meter and through the voltage meter to the aluminum foil, and the process continues.

A device that can run on the amount of electric current generated by the battery, such as an LED light, can be inserted in place of the voltage meter and will be powered by the electrical current from the battery. The difference in electrical potential (voltage) between the positively charged aluminum and the negatively charged penny creates the flow of electric force.

PHYSICS

V. Just For Fun

Steel Wool Meets Battery

What do you think will happen when steel wool meets a battery?

Materials

> 9 volt battery
> fine steel wool, plain (no soap)
> ovenproof pan or dish
> heatproof pad or surface

This experiment is best done outside. Have an adult help you.

❶ Take a piece of steel wool and spread it out with your fingers until it is a loose ball.

❷ Place the steel wool ball in an ovenproof pan that is on a heatproof pad or surface.

❸ Hold the 9 volt battery at the bottom and quickly and carefully touch the terminals to the steel wool. What happens?

❹ Record your observations.

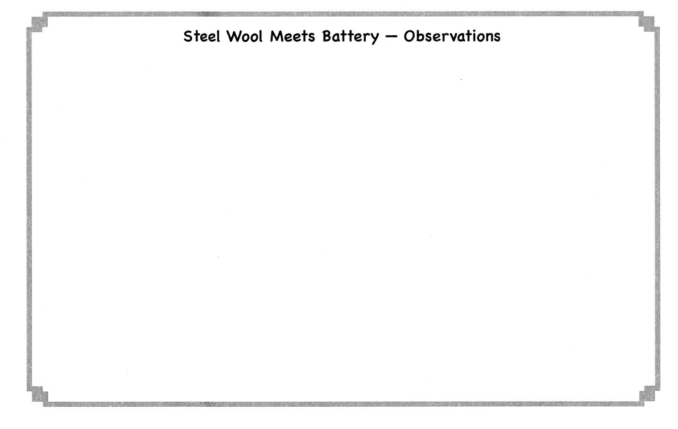

Steel Wool Meets Battery — Observations

PHYSICS

Why did what happened happen?

In the space below write your conclusions, then send a summary to us at: office@gravitaspublications.com.

If you're right, we'll send you a surprise gift!

PHYSICS

Steel Wool Meets Battery — Conclusions

Experiment 11

Charge It!

Introduction

Build your own instrument to detect electric charge—an electroscope!

I. Think About It

❶ What are the parts of an atom that have a charge and what charge do you think they carry? How do you think electric charge is important to an atom?

_____

_____

_____

_____

❷ What do you think happens when a positive charge meets a negative charge? Why?

_____

_____

_____

_____

❸ What do you think happens when a negative charge meets another negative charge? Why?

_____

_____

_____

_____

PHYSICS

❹ What do you think would happen if all the parts of an atom were positively charged? Why?

_____

_____

_____

_____

_____

❺ What do you think has happened when an object goes from having no charge to having a negative charge? From no charge to a positive charge? Why?

_____

_____

_____

_____

_____

❻ How would you explain electrical force to someone who doesn't know much about physics?

_____

_____

_____

_____

_____

PHYSICS

II. Experiment 11: Charge It!

Date _____

Objective _____

_____

Hypothesis _____

_____

_____

Materials

small glass jar with lid
aluminum foil
paperclip
duct tape (or other strong tape)
plastic or rubber rod (or balloon)
silk fabric (or your hair)
scissors
ruler
awl or other tool to make a hole

EXPERIMENT

❶ Cut two narrow strips of aluminum foil of equal length (about 2.5 cm [1 inch] long).

❷ Poke a small hole in the center of the lid of the glass jar.

❸ Bend open a paperclip to make a right angle from the outer loop of wire and a small hook from the inner loop. (See illustration on the right.)

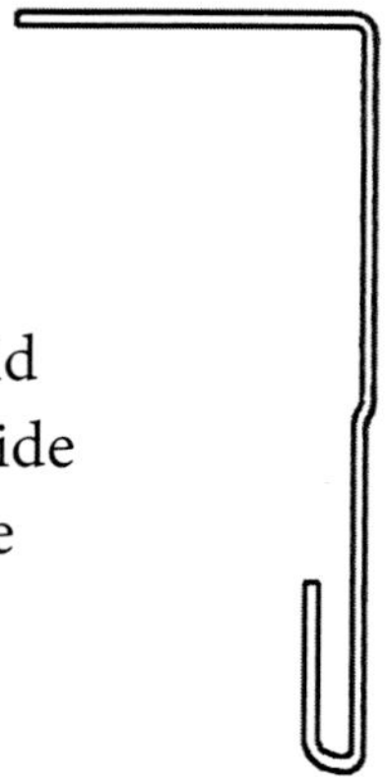

❹ Push the straight part of the paperclip through the small hole in the jar lid starting from the bottom side of the lid. Secure the paper clip to the outside of the lid with strong tape, leaving the end of the paper clip exposed. (See illustration on next page.)

❺ Hang the two strips of aluminum foil from the hook that is on the underside of the jar lid. Place the lid on the jar with the aluminum foil hanging from the hook inside the jar.

You now have an electroscope.

❻ Take the plastic or rubber rod and rub it with the silk fabric, or take the balloon and rub it in your hair or on the cat.

❼ Gently touch the rod or the balloon to the end of the paper clip that is on the outside of the jar lid.

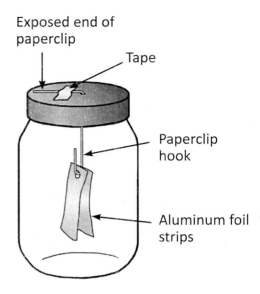

Exposed end of paperclip

Tape

Paperclip hook

Aluminum foil strips

❽ Observe the two pieces of aluminum foil and record your results.

Results

PHYSICS

III. Conclusions

What conclusions can you draw from your observations?

_____

_____

_____

_____

_____

_____

_____

_____

_____

_____

_____

_____

_____

_____

PHYSICS

IV. Why?

Static electricity is generated when an object like a balloon is rubbed on hair. Friction causes electrons to move from the hair to the balloon or the silk fabric to the rod, causing the rod or the balloon to be negatively charged. Electrons have a negative charge, so an object that gains electrons will become negatively charged.

The electroscope works with static electricity and the fact that like charges repel. In a charged object the like charges always move as far away from each other as they can. When you touch the paperclip in the electroscope with a charged object such as the balloon or plastic rod, the electrons move as far apart as possible, spreading from the rod along the paperclip and then along the aluminum strips. Both of the foil strips then have negative charge, so they repel and move away from each other.

After awhile the charge leaks away, and the foil strips come back together. The more charge, the more strongly the strips repel; therefore, you can tell how strongly charged the rod was when you started.

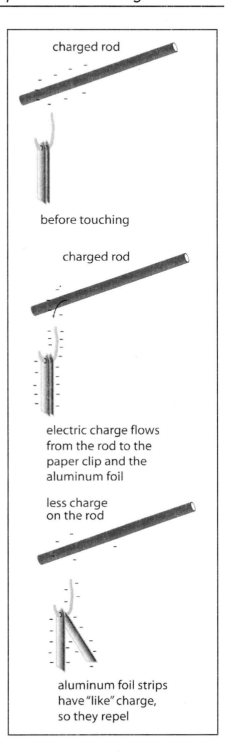

charged rod

before touching

charged rod

electric charge flows
from the rod to the
paper clip and the
aluminum foil

less charge
on the rod

aluminum foil strips
have "like" charge,
so they repel

PHYSICS

V. Just For Fun

More static electricity experiments!

Here are some quick experiments that show static electricity in action. For these to work well, you need to be in a room with low humidity. For each experiment, charge the straw or comb by rubbing it with a paper tissue or a cloth made of silk or wool. You may want to have a helper when you're charging two objects so you don't discharge one of them by touching it while you are charging the other.

Space is provided for you to record your observations.

Materials

several thin, bendable plastic straws
paper tissues (Kleenex) or cloth made of silk or wool
small piece of paper
small piece of aluminum foil
scissors
1 or more books
1 or 2 plastic combs
plastic cup
shallow bowl or a plate

⚡ Cut some very small pieces of paper and aluminum foil. What do you think will happen when you put a charged straw near the pieces? How close do you think the straw will have to get for something to happen?

⚡ Do you think you can use a charged straw to turn a page in a book?

⚡ Turn on a faucet so a thin stream of water is coming out. Charge a plastic comb. What do you think will happen as you move the charged comb slowly toward the water? What do you think will happen if you move the comb around or wiggle it? Do you think anything different will happen if you use two charged combs?

Charge a straw. Repeat the experiment. Do you notice any differences between how the water reacts to the straw and to the comb? To two charged straws? A comb and a straw?

⚡ Bend two straws slightly, then charge them. Hold one in each hand by the short part so the long parts are parallel and upright. What do you think will happen as you move them close together?

⚡ Charge two straws. Hold them so they are parallel to the floor and aligned one above the other and with your hands one above the other. What do you think will happen as you move the straws closer together? (Hold the top straw loosely—just enough to keep it aligned over the bottom one and not swinging from side to side.)

⚡ Take a shallow bowl or plate and put just enough water in it that a plastic cup will float. Charge one end of a straw and place the straw on top of the cup (across the rim). What do you think will happen if you charge another straw and bring it close to the straw that's on the cup? What will happen if you move the second straw around? Do you think if you used a charged comb, you would get different results? Two straws? Two combs? A straw and a comb? What if you charged the whole straw that's on the cup instead of just one end?

⚡ What other static electricity experiments can you make up?

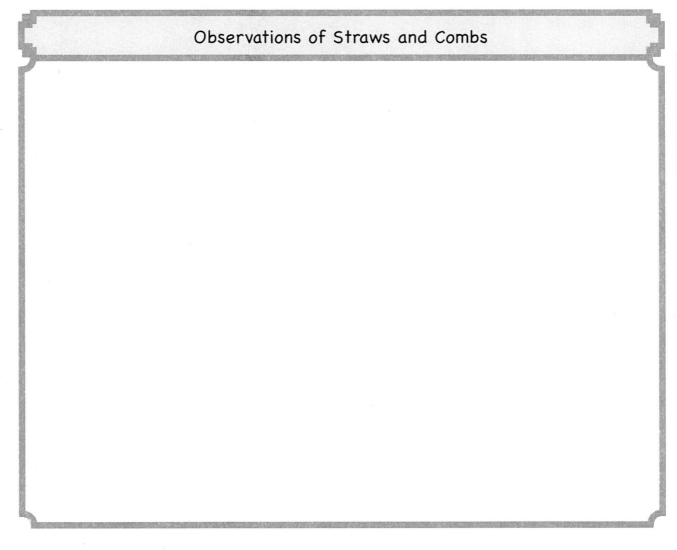

Observations of Straws and Combs

PHYSICS

More Observations of Straws and Combs

PHYSICS

Experiment 12

Circuits and Ohm's Law

Introduction

Explore electric circuits.

I. Think About It

❶　How do you think an electric circuit works?

_____

_____

_____

_____

_____

❷　What do you think is needed for an electric current to work?

_____

_____

_____

_____

_____

❸　How would you describe Ohm's Law in your own words?

_____

_____

_____

_____

_____

PHYSICS

❹ How much voltage do you think it would take to light a small light bulb? Why?

_____

_____

_____

_____

_____

_____

❺ How much current do you think it would take to light a small light bulb? Why?

_____

_____

_____

_____

_____

PHYSICS

❻ What do you think would happen if you put a resistor in your circuit? Would the light bulb still light? Why or why not?

_____

_____

_____

_____

_____

_____

II. Experiment 12: Circuits and Ohm's Law　Date _____

Objective _____

_____

Hypothesis _____

_____

_____

Materials　　　　　　　　　　　　　　　**Symbol**

(2) D cell batteries and battery holder　　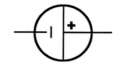　　battery

(1) 3.7 volt light bulb and socket　　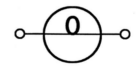　　light bulb

(1) switch　　　　switch

(4) alligator clip connectors　　　　wire

(2) 5 ohm, ¼ watt resistors　　　　resistor

(1) DC motor with propeller　　　　motor

EXPERIMENT

Part I: Building a simple circuit

❶ Using the symbols in the *Materials* section, draw a simple circuit. Include the battery source, the light bulb, and the switch.

❷ Build a circuit according to your diagram and close and open the switch. Record your observations below.

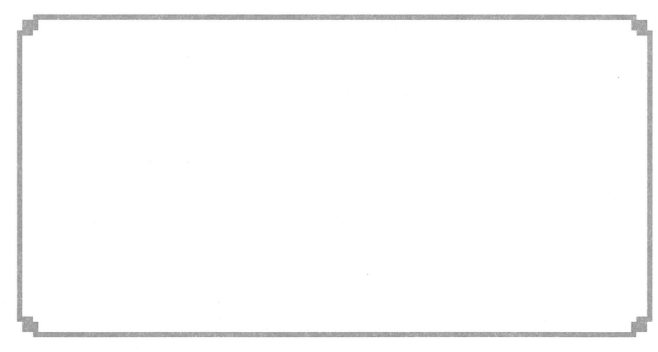

❸   Repeat Step ❶ and include the DC motor in your drawing of the circuit.

❹   Build your circuit, close and open the switch, and record your observations below.

Part II: Testing Ohm's Law

❶   Build the following circuit. Close the switch and note the brightness of the bulb. Record your observations.

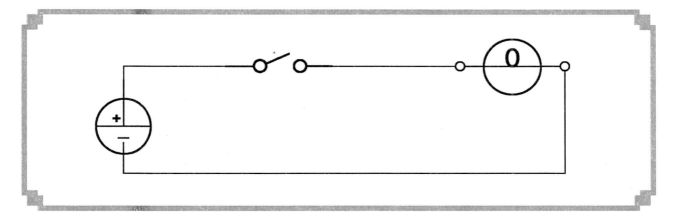

❷   Open the switch and insert a resistor between the light bulb and the battery. Draw your circuit.
Close the switch and observe the brightness of the light bulb.
Record your observations.

Drawing and Observations of Circuit with One Resistor

❸ Open the switch and insert a second resistor next to the first one. You may need to wrap or bend the end wires of the resistors together to make a good connection.
Draw your circuit.
Close the switch and observe the brightness of the light bulb.
Record your observations.

Drawing and Observations of Circuit with Two Resistors

III. Conclusions

Review your data. What conclusions can you draw from your observations? What does your experiment tell you about Ohm's Law?

_____

_____

_____

_____

_____

_____

_____

_____

_____

_____

_____

_____

_____

_____

_____

PHYSICS

IV. Why?

An electrical circuit is made when a power source is connected to a conducting wire and some type of load. A load is an electrical component or part of a circuit that draws, or consumes, electric power. A load could be a resistor, a motor, an electrical appliance, or a light bulb. In this experiment you created a simple circuit with all the elements (battery, conducting wire, and resistor) connected in a *serial circuit*. A serial circuit gets its name because all the components are connected in a *series* (one after another).

The electrical current in this experiment is created by chemical reactions inside the battery. When the switch is down, the circuit is *closed,* and electric current will flow from the battery through the conducting wires and through the filament in the light bulb, causing the bulb to light up. When the circuit is *open* (switch is open), the path of the electrical current through the wire is interrupted so no current can flow—no electrons can travel through the wire to illuminate the light bulb.

When a resistor is added to the circuit, the intensity of the light bulb is diminished because the resistor reduces the amount of electric current that can reach the bulb. Using Ohm's Law, you can see that as the resistance is increased, the current decreases.

Ohm's Law:

$$I = \frac{V}{R}$$

You can see from this equation, that as R (resistance) increases, I (current) decreases.

If two resistors are added to the circuit, there is enough resistance reducing the flow of electrons that the light bulb will not illuminate because there is not enough electric current getting to it.

PHYSICS

V. Just For Fun

Create a parallel circuit with two light bulbs. Draw your circuit and record your observations.

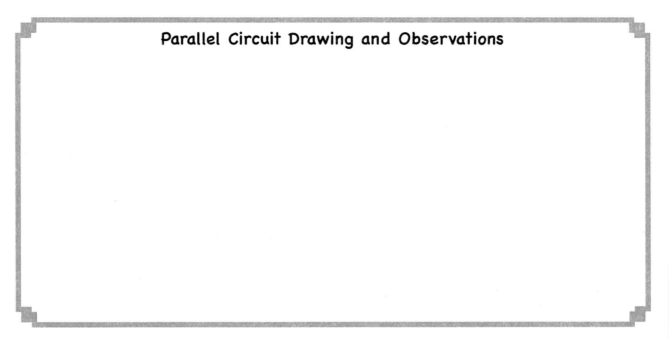

Parallel Circuit Drawing and Observations

Using the materials you have, build more circuits with your own variations. Draw the circuit and record your results for each variation. When you have finished the experiment, review your data and see what conclusions you can draw.

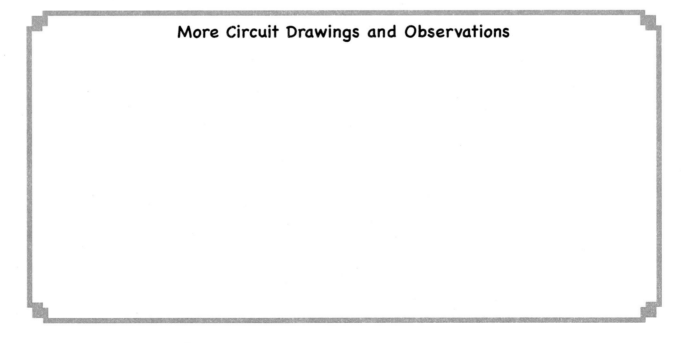

More Circuit Drawings and Observations

Even More Circuit Drawings and Observations

Conclusions

Review your data. What conclusions can you draw from your observations? What worked and didn't work? Why? What more did you learn about electrical circuits and Ohm's Law? What did you learn about serial and parallel circuits?

_____

_____

_____

_____

_____

_____

_____

_____

_____

_____

_____

_____

_____

_____

_____

_____

_____

PHYSICS

Wrap It Up!

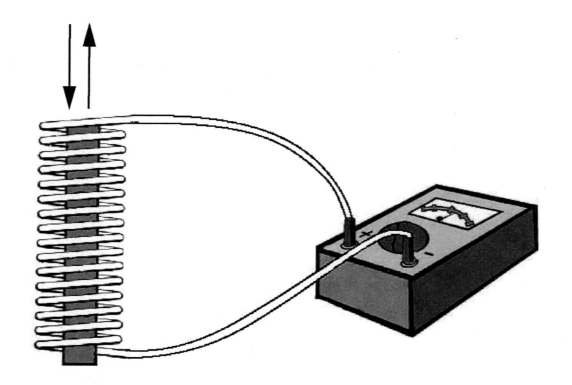

Introduction

Do you think you can make a magnet with a battery and some wire? Find out!

I. Think About It

❶ How would you find out if an object contains magnetic force?

_____

_____

_____

_____

❷ Do you think any material can be a magnet? Why or why not?

_____

_____

_____

_____

❸ What do you think would happen if you cut a magnet into little pieces? Why?

_____

_____

_____

_____

PHYSICS

❹   What happens if you place two magnets with their north poles together? Why?

_____

_____

_____

_____

_____

_____

❺   Do you think magnetic force can be useful? Why or why not?

_____

_____

_____

_____

_____

_____

❻   Do you think an electromagnet might have any advantages over a bar magnet? Why or why not?

_____

_____

_____

_____

_____

_____

PHYSICS

II. Experiment 13: Wrap It Up!   Date _____

Objective _____

_____

Hypothesis _____

_____

_____

Materials

    metal rod (a large nail or unmagnetized screwdriver can be used)
    electrical wire
    10-20 paperclips
    6v or larger battery
    electrical tape or 2 alligator clips
    scissors
    wire cutters
    Optional: wire stripping tool

EXPERIMENT

❶ Cut the electrical wire so that it is .3-.6 meter (1-2 feet) long.

❷ Trim the plastic coating off the ends of the wire so that there is about 6 mm (1/4 inch) of exposed metal on each end of the wire.

❸ Tape one end of the wire to the positive (+) terminal of the battery. (Alligator clips may be used in place of tape.)

❹ Tape (or clip) the other end of the wire to the negative (-) terminal of the battery.

❺ Take the metal rod and touch it to the paperclips. Record your results in the chart in the *Results* section.

❻ Coil the wire around the metal rod a few times. Both ends of the wire must remain hooked to the battery or be reattached if they come off.

❼ Touch the metal rod to the paperclips. Count the coils and record your results.

❽   Wrap another 1 to 5 coils around the metal rod.

❾   Touch the end of the metal rod to the paperclips. Record the number of coils and how many paper clips were picked up.

❿   Continue adding coils to the metal rod and counting the number of paperclips that can be picked up. Record the results each time you increase the number of coils. When you are finished, make a graph of your data.

Results

Number of Coils	Number of Paperclips

Graph your results

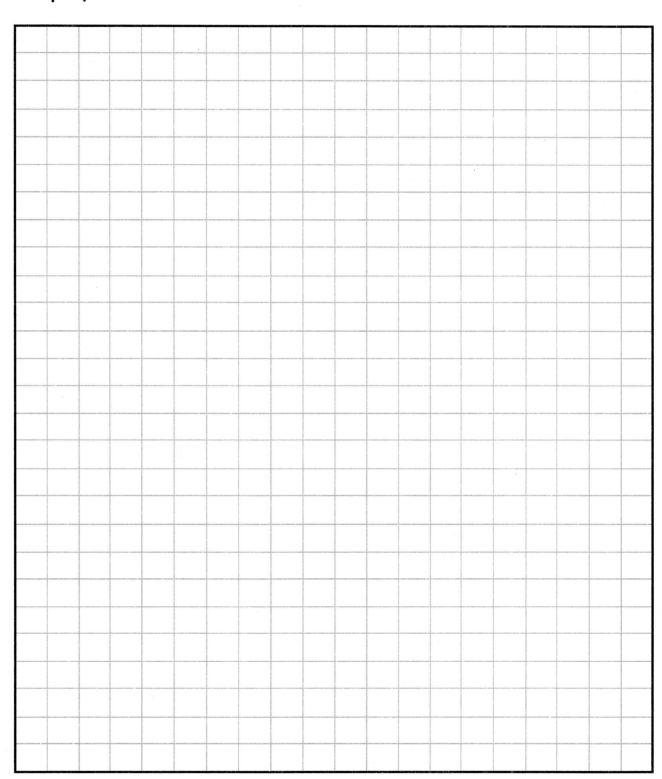

III. Conclusions

Review your data. What conclusions can you draw from your observations?

_____

_____

_____

_____

_____

_____

_____

_____

_____

_____

_____

_____

_____

PHYSICS

IV. Why?

An electromagnet is created when an electric current flows through a wire and causes a magnetic field that circles the wire. The more electric current that passes through the wire, the stronger the electromagnet will be — when there is more electric current, there is a stronger magnetic field.

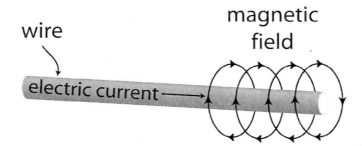

If the wire is wound into a coil, it can behave much like a bar magnet when the electric current is passed through it. The fields from each part of the coil add up to create a magnetic field that looks much like the field of a bar magnet.

In this experiment you observed that the electromagnet you created became stronger (picked up more paperclips) when you increased the number of loops around the metal rod and thus increased the strength of the magnetic field. Another way to increase the strength of the magnetic field would be to use a larger battery. You would notice that as the current in the wire increased with the stronger battery, a stronger electromagnet would be produced.

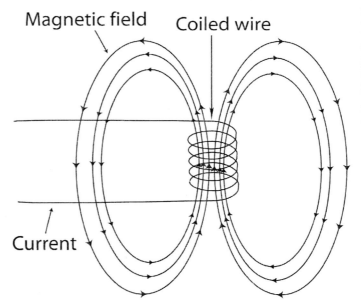

PHYSICS

V. Just For Fun

Seeing the Magnetic Field

Materials

bar magnet
plastic baggie
small flat-bottomed, clear plastic container with lid [about 5 cm x 8 cm x 1.5 cm (2" x 3" x 1/2") is a good size]
clear Karo syrup
spoon
2 pencils or other props
electromagnet from Wrap It Up! experiment

Experiment

❶ Take the bar magnet and place it in a plastic baggie. Take the bar magnet and the plastic container outside to collect iron filings from your yard or another place where there is exposed dirt. Hold the baggie containing the magnet and swirl it in the dirt. Small iron filings should collect on the outside of the baggie around the magnet.

❷ Carefully place the baggie in the plastic container and remove the magnet. The iron filings should fall off the baggie and into the box.

❸ Repeat Steps ❶-❷ several times until you have enough iron filings to thinly cover the bottom of the box.

❹ Pour enough Karo syrup into the box to just cover the iron filings and gently stir.

❺ Carefully place the plastic container on top of the electromagnet. Put props under the the container to lift it slightly above the electromagnet and keep it flat.

❻ Turn on the electromagnet and observe what happens over the next several minutes.

❼ Record your conclusions in the following space.

Conclusions

What conclusions can you draw from your observations?

What Makes an Aquifer?

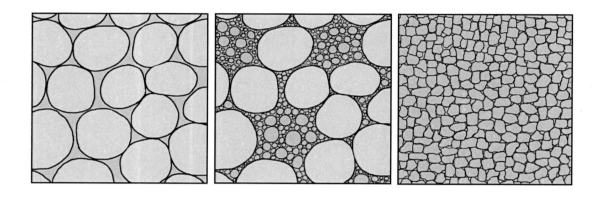

Introduction

See how water travels through some materials that make up Earth.

I. Think About It

❶ Do you think water can travel through soil? Why or why not?

_____

_____

_____

_____

_____

❷ Do you think water can travel through rock? Why or why not?

_____

_____

_____

_____

_____

❸ Where do you think the water you drink comes from? What steps are needed to get it to your house?

_____

_____

_____

_____

_____

GEOLOGY

❹   What do you think would happen if there were no groundwater but only surface water? Why?

_____

_____

_____

_____

_____

_____

❺   If you lived where there was a shortage of water, what things could you do to use less and help the water supply last longer?

_____

_____

_____

_____

_____

GEOLOGY

❻   Have you seen water pollution in your area? What actions do you think could be taken to help reduce water pollution and clean up polluted water?

_____

_____

_____

_____

_____

II. Experiment 14: What Makes an Aquifer? Date _____

Objective _____

_____

Hypothesis _____

_____

Materials

gravel, about 600 ml (2.5 cups)
sand, about 600 ml (2.5 cups)
dirt (soil), about 600 ml (2.5 cups)
pottery clay, about 600 ml (2.5 cups)
water
(4) Styrofoam cups, about 355 ml (12 ounce) size
(4) 16 oz. clear plastic cups, glasses, or other clear containers
pencil
marking pen
measuring cups
graduated cylinder, 100 ml
large bowl

Optional

stopwatch or clock with a second hand

EXPERIMENT

Part I

This part of the experiment is a test for permeability of different materials.

❶ Use the pencil to poke several holes in the bottom of each of four Styrofoam cups. Put the same number of holes in each cup and make all the holes about the same size.

❷ Put 250 ml (1 cup) of each material (sand, gravel, dirt, and clay) in its own cup. Label each cup with the material it contains.

GEOLOGY

❸   One at a time, hold each cup over a large bowl and pour 120 ml (4 ounces) of water into the cup. Note how long it takes the water to drain through each material. You can use a stopwatch or a timer with a second hand, or you can visually note how quickly or slowly the water runs through each material compared to the others.

❹   In the *Results* section, note how long it takes the water to go through each material and write down your observations of the properties of the material.

Results

Permeability Test	
Material	**Relative Permeability & Properties**

Part II

This part of the experiment is a test for porosity. Porosity describes the size and quantity of void, or empty, spaces between the particles of a material. The porosity of a material describes the quantity of liquid it can hold, and the permeability of the material determines the speed at which liquid can flow through the material. It is possible for a material to be porous but not permeable (impermeable) if the pores (void spaces) are not connected.

❶ Take four 16 oz. clear plastic cups, glasses, or other clear containers. Measure 350 ml (1.5 cup) of each material (sand, gravel, dirt, and clay) and put it in its own cup. Label each cup with the material it contains.

❷ Measure 100 ml of water in the graduated cylinder. Take one of the containers and slowly pour water into the material in it, trying to avoid making a depression in the top surface of the material. Pour in water until the water is up to the top of the material in the container. How much water is left in the graduated cylinder? Record your observations in the Results section.

❸ Repeat Step ❷ for the other three materials.

Results

Porosity Test	
Material	**Amount of water left in graduated cylinder**

GEOLOGY

Use the following chart to calculate the porosity of each material.

❶   Calculate how much water was held in the pores of the first material you tested. To get this number, subtract the amount of water that was left in the graduated cylinder from the 100 ml of water you started with.

❷   Divide the amount of water you were able to pour into the material by the total amount of material. Express the answer as a percentage.

For example: If you were able to add 90 ml of water to one of the materials, the calculation would be:

90 ml water ÷ 350 ml material = .2571 = 25.71% porosity.

❸   Repeat Steps ❶-❷ for the remaining three materials.

Porosity	
Amount of water held in pores	**Porosity percentage**

III. Conclusions

Which material was the most permeable?   _____

Which material was the least permeable?   _____

Which material was the most porous?   _____

Which material was the least porous?   _____

By doing this experiment, what did you learn about permeability and porosity? Do you think permeability and porosity are related? Why or why not? What other conclusions can you draw from your observations?

_____

_____

_____

_____

_____

_____

_____

_____

_____

_____

_____

GEOLOGY

IV. Why?

In this experiment you probably discovered that of the materials you tested clay is the least porous and the least permeable, and gravel is the most porous and the most permeable. Clay is made of very small particles packed closely together. This leaves very little pore space to hold water. Also, because the space between particles is so small, it is difficult for water to travel through clay because it is less permeable.

Gravel contains large particles that form larger spaces between them. This allows more water to enter the gravel. Because the pore spaces are so large, water flows rapidly through gravel.

Soils are a mixture of different materials. Soils that are mostly sand will allow water to run through rapidly, making them dry quickly, and they will not have many nutrients to support plant life. In a soil that is made mostly of clay, water does not move well through it, and the tiny pores in the clay can make it difficult for plant roots to push through. The best soil for growing plants has the right mixture of materials that allow the soil to retain enough water for the plants to use, drains well enough that plants don't get too much water, and has enough organic matter to provide nutrients for the plants.

From this experiment you can see that the materials an aquifer is made of will affect the quantity of water the aquifer can hold and how fast water can flow through it. Also, the type of soil above the aquifer will affect how fast the aquifer can recharge. A soil that is mostly clay won't allow much precipitation to infiltrate and a gravely soil will allow water to flow through quickly.

GEOLOGY

V. Just For Fun

Build aquifers!

❶   Aquifers hold and carry water under the ground. Look at the results of your permeability and porosity tests and think about how you could layer the materials you tested to make an effective aquifer.

❷   Make a trough for building your aquifer. You can fold a piece of cardboard into a U shape and cover it with plastic. First decide how long, wide, and deep you want the aquifer to be and then have an adult help you cut the cardboard accordingly. Or you may have another idea about how you can build a trough.

❸ There are different types of aquifers, including those that hold the water in one area, those that let it flow short distances, and those that carry it long distances. The rate at which water flows also varies. Think about how you might use the materials you tested to create different types of aquifers.

❹ Use fresh samples of the materials for this part of the experiment. Choose two or more of the materials you tested and layer them in the trough. You may want to put screening or coarse cloth over the ends of the trough to keep the materials from washing out. This is your aquifer.

In the chart in the Results section, list the materials in the order in which they are layered in each aquifer.

❺ You may want to do this experiment outdoors or you can hold your aquifer at a slight angle with the lower end over a bucket. Pour water slowly on the higher end of your aquifer and note what happens. You can also try tilting the aquifer at a greater angle and pouring more water in the top end. What else could you try?

Record your observations in the *Results* section.

❻ Repeat Steps ❹-❺ to make several more aquifers. You can also add different materials of your choice to your aquifer.

Results

In the following charts record information about the layers in each of your aquifers and your observations about water flow, aquifer tilt, and any other variations.

When you have finished the experiment, review your data and write a summary of your observations. Then review your summary and based on these observations write conclusions about what you learned from making model aquifers. Also record how you think your model aquifers compare to real aquifers.

Aquifer

Aquifer Layers (Materials in order from top to bottom)	Observations

Aquifer

Aquifer Layers (Materials in order from top to bottom)	Observations

Aquifer

Aquifer Layers (In order from top to bottom)	Observations

Summary of Aquifer Observations

_____

_____

_____

_____

_____

_____

_____

_____

_____

_____

_____

_____

_____

_____

_____

_____

_____

GEOLOGY

Conclusions

_____

_____

_____

_____

_____

_____

_____

_____

_____

_____

_____

_____

_____

_____

_____

GEOLOGY

Experiment 15

My Biome

Introduction

Take a close look at the biome in which you live. What features make it your biome?

I. Think About It

❶ Do you think that you often pay attention to everything around you when you go outside? Why or why not?

_____

_____

_____

_____

❷ What do you think the area where you live would be like if there were no insects? Why?

_____

_____

_____

_____

❸ If the weather patterns during the course of several years became totally different from what they are now, do you think you would still be living in the same type of biome?

_____

_____

_____

_____

GEOLOGY

❹   Do you think animals behave in the same way every day of the year? Why or why not?

_____

_____

_____

_____

_____

❺   Do you think there is a relationship between the kinds of animals and the kinds of plants found in your biome? Why or why not?

_____

_____

_____

_____

_____

❻   How do you think the land formations and soils affect the plant and animal life in your biome?

_____

_____

_____

_____

_____

GEOLOGY

II. Experiment 15: My Biome   Date _____

Objective _____

_____

Hypothesis _____

_____

Materials

field notebook
pencil and colored pencils
small backpack
water bottle
snack

Optional

binoculars

EXPERIMENT

❶ Pack a small backpack with a water bottle, pencils, your field notebook, and a snack.

❷ Take a one to two hour hike in your surrounding environment.

❸ In your field notebook, record what you see. Pay attention to everything around you. Carefully observe the plant and animal life, where organisms live, and what they are doing. Observe the landforms, rocks, and soil and how these affect what lives in an area.

❹ Describe the types of plants in your surroundings and the types of animals (both large and small) and observe their interactions. Are ants crawling on plants? Are insects and animals eating plants? Are there dogs playing in the grass? Are cats looking for birds or insects to catch? Are birds pulling worms out of the ground? Are plants blooming?

❺ Observe the weather and how the weather affects the plant and animal life in your surroundings. Is there snow on the ground and do the plants look lifeless? Is it warm and sunny with lots of flowering plants? Are there insects out foraging for food? Are lizards sunning themselves? Is it raining? Or windy? Are the birds cheeping or sitting quietly waiting for a storm to pass? Record your observations in your field notebook.

❻ After your hike, review your notes and record your observations in the chart provided in the *Results* section.

GEOLOGY

Results

Summarize your observations below.

Summary of Biome Observations	
Route followed	
Weather	
Plants	
Animals	
Interactions	

Summary of Biome Observations

New things, interesting things, surprising things, and more!

III. Conclusions

Use the chart in the *Student Textbook* to determine the type of biome you live in. Record the name of your biome, use your observations to describe it, and list any unique or surprising features you discovered. Think about which features you would include if you were to describe your biome to someone who has never visited you.

_____

_____

_____

_____

_____

_____

_____

_____

_____

_____

_____

_____

_____

_____

_____

IV. Why?

An environment is the set of conditions surrounding an organism in the region where it lives. An ecosystem is a specific area that contains a community of living things existing under similar conditions. An ecosystem can be any size—from very small to very large. On the other hand, a biome is always a very large region and is an ecosystem that is defined by the climate, soils, and plant life that exist within it.

By exploring your biome in this experiment, you were able to observe different plants and animals and the conditions under which they live. You probably noticed large animals such as dogs and people and very small animals such as insects; plants as large as trees and very small plants such as grass. You may have noticed what conditions different plants require to grow (how much sunlight, how much water, what type of soil) and whether some plants were being eaten by bugs or animals. You may have found a home that an animal built, such as a burrow or a bird's nest. With careful observation, you can begin to see how the different spheres of Earth interact to create the biome and how a change in certain conditions in the biome, such as the amount of rain, could affect the organisms that live there.

Think about what you observed on your hike and take another look at your notes. Did you describe different ecosystems within the biome? Did you pass a marsh, a pond, or other wet area? A meadow and a forest? Farmed land and front yards?

If we use the example of a temperate deciduous forest biome, we can see that although they are part of the same biome, a pond and a wooded area would be different ecosystems containing different communities of plant and animal life. Some plants and animals require the watery ecosystem of a pond, while others need an area that has a thick growth of trees. Within these different ecosystems different habitats will be found. A fish in the pond would be in a habitat with the right features to allow it to live; for example, the right kind of food, enough space, deep enough water (or perhaps shallow water), and a way to hide or escape from predators. The pond might have the right conditions for a water lily habitat and another habitat for a certain kind of frog to thrive. On the other hand, the wooded ecosystem would provide a place for very different types of habitats. Shade-loving plants could find a place to grow beneath the trees. Deer might live there, finding the plants they need to eat, places to sleep, and protection from predators. The trees could house many types of birds.

We can see that a biome is a very large area that includes many different ecosystems with each ecosystem containing many different habitats. Figuring out the details of where an organism lives can get quite complicated!

GEOLOGY

V. Just For Fun

What's that bird?

Watching and identifying the birds around you is an interesting and fun way to learn more about ecosystems and habitats (and birds!). And you can bird-watch anywhere—even when you're riding in a car. Here are some suggestions for getting started.

- Get a field guide to the birds book that has good pictures and descriptions of the birds in your area. Many birders like the Peterson field guides, but other books are also good. Spend some time leafing through the book and familiarizing yourself with different types of birds. Which ones do you think you have seen in your area? When you go outside or look out your window, pay attention to the birds you see and then find them in the field guide. What can you learn about them? Binoculars of any kind are helpful when bird-watching.

- If you have a good location (and no outdoor cats), set up one or more bird feeders with different kinds of seeds and maybe a feeder for hummingbirds too. Tend the feeder daily and observe the birds. Do different birds feed at different times of day? Do you only see them at a certain time of year? Do some birds chase other birds away and hog all the food? Use your field guide to see how many birds you can identify. Even without a feeder, you can watch and identify the birds in your yard and neighborhood.

- Gather up your hiking gear (backpack, binoculars, field guide to the birds book, field notebook, pencils, water, and snack) and go for a walk anywhere outside. When you spot a bird, see if you can identify it. Make notes about what kind of bird it is, where and when you saw it, and its color and markings. Also note anything else of interest. What is it doing? What is it eating? What kind of habitat is it in? How many other birds of the same kind are with it? What else can you notice? If the bird sits still long enough and you have a camera, you can photograph it. Or you can make a sketch of it.

- Try pishing. Stand still and make a sound like "psshh, psshh, psshh." Curious birds may gather near you to see what's going on. A little patience may be required.

- If you have a smart phone or tablet, you can get a bird identification app. The Audubon Society has a free app that has bird calls as well as visual identification information (http://www.audubon.org/apps). Another app is called *Peterson Birds — A Field Guide to Birds of North America* and is available for a small fee. With a little research you'll be able to find other apps too. Note that some apps may work only where you have cellphone or wifi connection.

GEOLOGY

Experiment 16

Finding North

Introduction

Make your own compass and use it to find north!

I. Think About It

❶ How do you think magnets can be useful?

_____

_____

_____

_____

❷ How do you think you can tell whether or not Earth has a geomagnetic field?

_____

_____

_____

_____

❸ Do you think Earth has any features that are similar to a magnet? If so, what are they?

_____

_____

_____

_____

❹   If you were exploring the Arctic, do you think you could find the north end of Earth's axis by using a compass? Why or why not?

_____

_____

_____

_____

_____

_____

❺   How do you think Earth's geomagnetic field and the magnetic field of an electromagnet are similar? How do you think they're different?

_____

_____

_____

_____

_____

GEOLOGY

❻   If there were no magnetosphere, do you think life on Earth would be different? Why or why not?

_____

_____

_____

_____

_____

II. Experiment 16: Finding North   Date _____

Objective _____

_____

Hypothesis _____

_____

_____

Materials

steel needle
bar magnet
piece of cork
tape
medium size bowl
water

EXPERIMENT

❶ Take the bar magnet and slowly stroke the needle against it for about 45 seconds. This will magnetize the needle.

❷ Find an object that a magnet will stick to and test the needle to see if it will stick to that object. If the needle doesn't stick, rub it against the bar magnet for a while more, and then test it again.

❸ Center the magnetized needle on the top surface of the piece of cork and tape it in place.

❹ Pour water into the bowl until it is almost full, and carefully place the cork in the center of the bowl so it is floating, needle-side up. The needle should not be touching the side of the bowl.

❺ Observe what happens. Can you tell which way is north? Why or why not?

❻ In the following box, note your observations.

Observations

❼   In the box in the *Results* section, draw a simple map of the room you are in. Indicate the walls, doorway, and any windows.

❽   Place your map on the table next to your cork and needle compass. Turn the map so that it is in the same orientation as the room. (Match the direction of the walls on your map to the walls in the room.)

In the middle of the map, draw a line that goes in the same direction as the needle of your compass.

❾   Think about the approximate locations of sunrise and sunset in relation to the room you're in. Knowing that the sun rises in the east and sets in the west, mark the approximate locations of east (E) and west (W) on your map.

❿   Can you now tell which end of the needle is pointing north? If so, mark it on your map with the symbol N.

Results

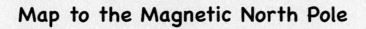

Map to the Magnetic North Pole

III. Conclusions

Based on your observations, what conclusions can you make about the ease or difficulty of making a compass and finding north? What other conclusions can you draw from your observations?

_____

_____

_____

_____

_____

_____

_____

_____

_____

_____

_____

_____

_____

_____

GEOLOGY

IV. Why?

In this experiment when you rubbed the steel needle with the bar magnet, you induced a magnetic force that magnetized the needle. Recall from Chapter 13 that a magnet placed close to an object made of a material such as iron can cause that object to become a magnet by induction.

Like a bar magnet, Earth's magnetic field has a north pole and a south pole, and thus Earth is said to have a dipole magnetic field. *Di* is from Greek and means "two." The needle of a compass points north because the magnetized metallic needle is attracted to Earth's North Magnetic Pole. Earth's magnetic field is referred to as the geomagnetic field, and it surrounds the Earth, going outward from the core at the Magnetic South Pole and in at the Magnetic North Pole.

Scientists think that Earth's magnetic field is electromagnetic. As you discovered in Experiment 13, an electromagnet can be built by running an electric current through a wire that is coiled around an iron rod.

Although there aren't wires in Earth's core, the geomagnetic field is thought to work in a way that is similar to an electromagnet with the geomagnetic field being generated deep within Earth in the liquid outer core. Electrical forces are created in the outer core due to convection currents that are caused by variations in the temperature, pressure, and composition of molten iron and nickel. In addition to convection currents, the spin of the Earth causes a spiralling, or coiling, motion in the molten materials. This motion is called the Coriolis force which also aligns the spirals (coils) of molten materials into a north/south orientation. The resulting electrical forces work as an electromagnet, creating the geomagnetic field. Once you induced magnetic force in the needle in this experiment and allowed it to turn freely, the needle was able to "find" the North Magnetic Pole.

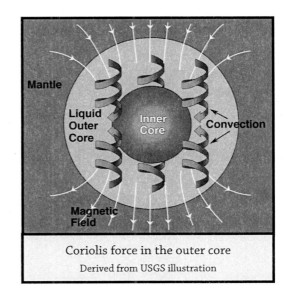

Coriolis force in the outer core

Derived from USGS illustration

V. Just For Fun

Finding treasure!

In this experiment you'll use a compass to make a map and have your friend use the map and compass to find a hidden treasure.

Materials

small object of your choice for the treasure
compass

Experiment

❶ Practice finding North (N), South (S), East (E), and West (W) with your compass. The needle will always be pointing to North. Stand with the compass in front of you, flat in your hand and parallel to the ground. Hold the compass in the same position as you turn in different directions.

Turn your body until the needle lines up with N. You are now facing North.

Now turn to the right until your body is lined up with the E on the compass. You are now facing East and the needle will be pointing to your left.

Turn to the right again until you are lined up with the S on the compass. You are now facing South and the needle will be pointing in the opposite direction, behind you.

Turn again to your right until you are lined up with the W on the compass. You are now facing West and the needle will be pointing to your right.

North, South, East, and West are called the four cardinal directions. Each is 90° from the previous one, or 1/4 of a 360° circle. If your compass has degrees shown on its face, the directions will be: N 0° , E 90°, S 180°, and W 270°.

❷ This treasure hunt can be done outdoors or indoors. A box is provided in the next section where you can draw your map. First draw an outline of the area where the treasure hunt will take place. Find North and mark it on your map with an arrow in the proper orientation relative to the outline. If you wish, you can add some features such as trees, chairs, etc.

❸ Select a place where you would like to hide the treasure.

❹ Pick a starting location. Place an object here or make a mark at this spot. Put the starting point on your map.

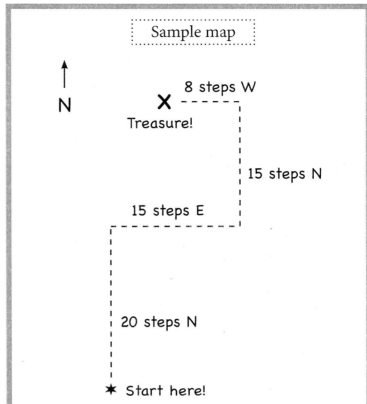

Sample map

N

X Treasure! — 8 steps W

15 steps N

15 steps E

20 steps N

★ Start here!

❺ Use the compass and the cardinal directions to chart a path to the treasure. On the map write the direction of travel and how many heel-to-toe steps are to be taken between each change of direction. (You can chart your map first and hide the treasure at the end of your route.)

❻ Give the map to a friend and have them use just the map and the compass to find the treasure. You can play this game more than once, refining your map or making the map a bit more complicated each time. Your friend can try making a map for you to follow.

Treasure Map

GEOLOGY

Record your observations. How well did your map work? Did you have any problems making or using it? If so, how did you solve them? What did you notice about using a compass?

Observations

Experiment 17

Solve One Problem

Cleaning up trash along the Anacostia River, Washington, DC

Courtesy of Gwen Bausmith, US Environmental Protection Agency (US EPA)

Introduction

The solution to a problem starts with a thought!

I. Think About It

❶   Do you think there's a way that Earth's spheres could operate separately from one another? Why or why not?

_____

_____

_____

_____

❷   Which of Earth's spheres do you think might be affected by a big blizzard? How and why?

_____

_____

_____

_____

❸   If you were an Earth system scientist, how do you think you might look at waste water disposal differently than a chemist? Why?

_____

_____

_____

_____

❹　If there is an oil spill in the ocean, which different scientific disciplines do you think might help solve the problem? How?

_____

_____

_____

_____

❺　If you needed to solve an environmental problem in your community, how do you think you might go about looking for new ideas?

_____

_____

_____

_____

❻　What do you think is being done in your community now to help keep Earth's spheres healthy and in balance?

_____

_____

_____

_____

GEOLOGY

II. Experiment 17: Solve One Problem   Date _____

Objective   _____

_____

Hypothesis   _____

_____

_____

Materials

pencil, pen
imagination

Optional

notebook

EXPERIMENT

❶   Spend a day observing your neighborhood. Go outside and watch how the people in
your neighborhood interact with their surroundings. Observe how people's interactions
affect other people's lives, plant life, and other animal life.

❷   Make a list of all the problems you discover. Include any details you notice. For
example, maybe your neighbor is elderly and can't get her trash to the curb, so it piles
up, spills over, and then the dogs eat it and scatter it across the street and get sick.
Or maybe your neighbors walk their dog but never pick up their dog's waste and,
unhappily, other people step in it, tracking it into the local store where it contaminates a
cantaloupe that has fallen on the floor on top of it.

❸   Record how many different spheres these problems influence. For example, trash in
the yard interacts with: Earth's crust (trash in the dirt), the biosphere (animals eat the
trash), the atmosphere (the trash gives off an odor), and possibly the hydrosphere (if it
rains and the water becomes contaminated).

GEOLOGY

Observations of Problems

Problem	Spheres Influenced

GEOLOGY

❹   Examine the problems and imagine a solution to each problem that you could actually carry out. For example, if your neighbor can't get the trash to the curb in time for the trash pickup, then maybe you could volunteer to move the trash for her. List your solutions below.

Solutions to Problems

Problem	Solution

❺ Now that you have solved the problems, explore how you might expand your solutions to solve the problems for a larger number of people. For example, perhaps you can recruit 5 friends to volunteer so 6 elderly people will have help moving their trash to the curb every week. List how you could expand your solutions to serve more people.

Expanded Solutions

Problem	Solution

GEOLOGY

❻ Now that you have found a way to expand your solutions to serve a greater number of people, explore how you could turn one of your ideas into a business that would allow you to generate resources and jobs for others while helping to solve more problems. For example, you could start a nonprofit company and get donations from the city government to help recruit more volunteers to help more elderly with their trash. Describe below how your idea could become a business or nonprofit organization.

Results

Summarize your results below.

Results	
Questions	**Ideas**
What was the problem?	
Which of Earth's spheres did the problem influence?	
What was the solution?	
How could you expand the solution to serve more people?	
How could you turn your solution into a business or nonprofit?	
When will you start implementing your solution in real life?	

GEOLOGY

III. Conclusions

It is said that knowledge is power. Describe below how understanding the science of the Earth — how it is put together and how it works — can help give you the power to solve real-life problems.

_____

_____

_____

_____

_____

_____

_____

_____

_____

_____

_____

_____

_____

_____

IV. Why?

In this experiment you explored how the principles of Earth system science can be applied to solving a problem that exists in your community. By observing how changes in one of Earth's spheres cause changes in the other spheres, we can better understand how human activities affect the balance of the spheres. With this understanding, we can begin to solve problems such as pollution or trash that needs to be put out to be collected.

Careful observation and the collection of data are necessary. What is the problem that is occurring? What is causing the problem? What are the effects of the problem? Once the problem, its causes, and its effects have been identified and analyzed, we can begin to think of solutions. The solutions, in turn, will be analyzed to see what effect they will most likely have on the other spheres. Since the interactions between the spheres is so complicated, outcomes can be uncertain, but analyzing problems and solutions before acting leads to better results.

You may have found that, in a similar way to Earth's spheres, communities can also be thought of as having spheres. Individuals are part of families, families are part of neighborhoods, neighborhoods are part of a larger community, and so on. There are small organizations such as Girl Scouts and Boy Scouts that are part of the national community of scouts. Local governments are part of state governments which are part of national governments. Each part has its own characteristics and actions, and all the parts, large and small, are interrelated and affect each other. Understanding how different groups interact and affect each other is an important step in problem solving.

Because groups and events are so interrelated, it's quite possible that the positive effects of coming up with one seemingly small solution to a problem will spread and have a very positive effect on the larger community.

GEOLOGY

V. Just For Fun

Now that you have come up with an action for solving a problem, think about how you would let others know about the problem and your idea for a solution. For example, you might create a flyer about the service you are providing for putting out the trash in your neighborhood and then hand out the flyer to your neighbors. You might write a newspaper article or a letter to the editor of your local newspaper telling people about your project, or you might start a blog. If you have a local radio or TV station, you might contact them about doing a news story. You might also make an informational video describing your project and send it to interested people.

In the space below, write your ideas for one or more actions you could take and how you would go about doing them.

_____

_____

_____

_____

_____

_____

_____

_____

_____

_____

_____

_____

GEOLOGY

Experiment 18

Astronomy Online

Courtesy of Gemini Observatory/AURA

Introduction

Access to astronomy data has expanded significantly because of the internet, allowing amateur astronomers to use online resources to explore sophisticated data. Here you will search the internet to find online tools for experiments in two following chapters.

I. Think About It

❶ What types of online programs have you used to explore astronomy?

_____

_____

_____

_____

❷ Have you ever used a star chart or online planetarium to look at planets in our solar system? If so, describe what you discovered.

_____

_____

_____

_____

❸ What other types of online resources have you used to explore any aspect of science? (For example, online chemistry charts, biology videos, physics demonstrations)

_____

_____

_____

_____

ASTRONOMY

❹   How easy or difficult has it been to use online resources to learn about science?

_____

_____

_____

_____

_____

❺   How reliable do you think the resources you've used are? How do you know the information you are viewing online is correct?

_____

_____

_____

_____

_____

_____

❻   Do you use reviews or forums to discover what other resources people are using to learn about science online? Why or why not?

_____

_____

_____

_____

_____

_____

ASTRONOMY

II. Experiment 18: Astronomy Online   Date _____

Objective   _____

_____

Materials

computer
internet access

EXPERIMENT

❶   Read through the experiments for Chapters 19 and 21. You will need to collect specific information for these experiments. Make a short list of the information you will need to collect (e.g., nebulae, galaxies, galactic center, globular clusters, constellations, etc.).

_____

_____

_____

_____

_____

_____

_____

_____

_____

_____

❷   Do an online search for astronomy software and websites. Try different keywords to find different types of resources, for example, "online astronomy," "astronomy programs," "astronomy software," etc. When you find a resource you might like to use, determine whether it will work on your computer. List the URLs for 3-5 resources along with a brief description of the type of information each provides.

_____

_____

_____

_____

_____

_____

_____

_____

_____

_____

_____

_____

_____

_____

_____

ASTRONOMY

❸ Review the resources listed in Step ❷. Which ones provide the information you are looking for? Do you need to download a program or can you use the program without downloading? Is it easy to use and understand? Does it allow you to search for the information you need, such as nebulae, galaxies, and the galactic center? Are the images easy to see? In the space below, make notes about the resources you've selected, including the types of information the program will provide for the remaining astronomy experiments.

_____

_____

_____

_____

_____

_____

_____

_____

_____

_____

_____

_____

_____

ASTRONOMY

Results

Briefly describe the online resource(s) you have selected and why you think it will work for the astronomy experiments in Chapters 19 and 21.

Astronomy Resource

_____

_____

_____

_____

_____

_____

_____

_____

_____

_____

_____

_____

_____

_____

_____

ASTRONOMY

III. Conclusion

What conclusions can you draw from your reserach?

_____

_____

_____

_____

_____

_____

_____

_____

_____

_____

_____

_____

_____

_____

_____

_____

IV. Why?

The internet can be a valuable resource for learning about astronomy. There are thousands of sites and programs available that contain star charts, images, and videos in addition to text. An amateur astronomer can find lots of fascinating information at levels everywhere from beginning to advanced.

In this experiment you researched astronomy software that can be used to find information for upcoming experiments. You discovered that there are several things that need to be considered when deciding upon an astronomy program to use for doing research. Can the program provide you with the data that will be needed? Is your computer compatible with the program's operating system requirements? Is the program understandable and easy to use? All of these factors play a role in determining the type of software you will need.

However, not all online resources are useful, and learning how to research science information is an important skill for using the internet effectively. For example, some sites require a user to download a program and install it on their computer. Some software requires a particular operating system, and if the computer you are using does not meet the program's requirements, the software cannot be used. Some software may be available only for purchase, which can be a limiting factor.

You may have discovered several online resources that seem like they would meet your needs, and you can use more than one resource to gather the data needed for the experiments that follow. In fact, it is good scientific practice to gather information from more than one resource whenever possible. Using more than one resource allows you to verify the information you are collecting. If you find inconsistencies between resources, you can try to determine which is the more reliable source, or you can check additional sources to see if the majority are in agreement about a fact. Because scientific study is complicated and scientists have differing theories and different experimental results, you may find that not everyone is in agreement. In looking for reliable resources, those that come from university sites are generally very reliable, as are sites that come from government agencies such as NASA, NOAA, and ESA (European Space Agency). Scientific journals can also be a good resource; however, care must be taken to ensure that these are fact-based publications with articles backed up by actual scientific study and data rather than just opinions. ·

The fun of doing internet research in astronomy is that it can lead you to the discovery of many amazing things in space that you didn't know existed. And with each new satellite, space telescope, space probe, lander, or rover that is launched, more and more discoveries are made that you can read about and see as images.

V. Just For Fun

NASA has many websites for its different missions and areas of research. Spend some time exploring various NASA sites—reading about NASA's missions and looking at the images and videos in the multimedia galleries. In the space provided make notes about the most interesting facts you find. Think about what information you would have to record in order to be able to quickly find a particular web page again and include this information in your notes. Also check out ESA websites. Do you find different information by going to space agencies from different countries?

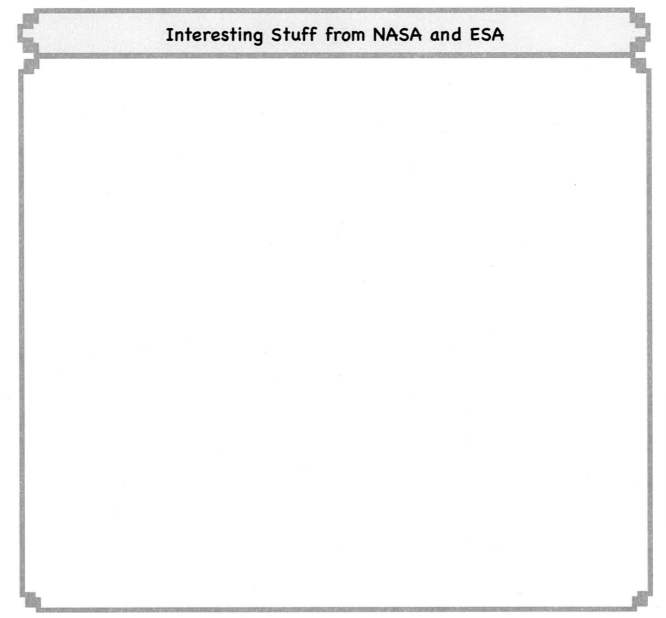

Interesting Stuff from NASA and ESA

ASTRONOMY

More Interesting Stuff from NASA and ESA

ASTRONOMY

Experiment 19

The Center of the Milky Way

Artist's concept courtesy of ESO/NASA/JPL-Caltech/M. Kornmesser/R. Hurt

Introduction

Locate the center of the Milky Way galaxy using globular cluster data.

I. Think About It

❶ How easy or difficult do you think it would be to find the center of the Milky Way Galaxy?

_____

_____

_____

_____

❷ How would you describe a globular cluster?

_____

_____

_____

_____

❸ How do you think knowing about globular clusters could be helpful in finding the center of the Milky Way Galaxy?

_____

_____

_____

_____

ASTRONOMY

❹ How many stars do you think a globular cluster contains? Why?

_____

_____

_____

_____

_____

❺ What objects do you think are located in the galactic center of the Milky Way? How would you describe them?

_____

_____

_____

_____

_____

❻ Do you think the galactic center can be observed from Earth? Why or why not?

_____

_____

_____

_____

_____

ASTRONOMY

II. Experiment 19: The Center of the Milky Way   Date _____

Objective   _____

_____

Hypothesis   _____

_____

Materials

computer
internet access

EXPERIMENT

❶   Set up the online resource you chose in Experiment 18.

❷   The *Appendix* at the back of this book gives data for globular clusters observed in our Milky Way Galaxy. The data table shows 158 globular clusters compiled as of June 30, 2010. From left to right the table lists the ID, name, and cross-reference for the cluster followed by the constellation where the cluster is located and various astronomical parameters associated with the cluster.

Look at the data table in the Appendix, and locate the three constellations that have the highest number of globular clusters. [Note: The number of globular clusters observed in a constellation is found in parentheses next to the constellation name. Constellations with fewer than two globular clusters are not listed.]

❸   In the chart below, record the three constellations that have the most globular clusters.

Constellation Name	# of Globular Clusters

ASTRONOMY

Results

Open the resource you have chosen for finding the galactic center of the Milky Way Galaxy. Search for the three constellations listed in Step ❸ of the experiment. If your resource shows their locations, record this information. Since globular clusters are most numerous in the galactic center, these three constellations will lead you to the center of the Milky Way. Do a search for the location of the galactic center to check your results. In the space below, record your observations.

_____

_____

_____

_____

_____

_____

_____

_____

_____

_____

_____

_____

_____

_____

_____

_____

ASTRONOMY

III. Conclusion

What conclusions can you draw from your research? Based on your observations, where is the galactic center of the Milky Way? How easy or difficult do you think it is to find the center of a galaxy?

_____

_____

_____

_____

_____

_____

_____

_____

_____

_____

_____

_____

_____

_____

ASTRONOMY

IV. Why?

In this experiment you used and evaluated data on globular clusters to determine the center of the Milky Way galaxy. The appendix at the end of this workbook lists a number of different constellations together with the identification numbers, distance from the Sun or galactic center, apparent magnitude, and apparent dimension. The number of globular clusters in a constellation is found in parentheses next to the constellation name. Using this information you should have discovered that the globular clusters that contain the highest number of stars are Sagittarius with 34 globular clusters, Ophiuchus with 25 globular clusters, and Scorpio with 20 globular clusters. Because we know that the densest group of stars is at the center of the Milky Way Galaxy, we can use these three constellations with the largest number of globular clusters to find the galactic center.

Learning how to read and sort through scientific data is an important skill. Scientists often have to work with large amounts of data, sorting through numbers, names, and symbols. It takes time to learn how to study and evaluate scientific data. In this experiment the data was presented in a chart with the globular cluster count already identified. It might have taken longer or been more difficult if this information was not presented on a chart. You also may have noticed that some of the information listed on the chart was not used to determine the galactic center.

Depending on the software you selected, you should have been able to verify your results. You should have been directed to the identical location searching on the words "galactic center" as you found by typing in the three constellations, Sagittarius, Ophiuchus, and Scorpio. Being able to verify a result is an important step for any scientific research. If you were not able to verify the galactic center, you can redo the experiment or use a different software program. If the programs you are using are reliable, you should be able to verify your results.

The globular cluster chart includes information about right ascension, declination, and dimensions in arc minutes. With a little further research you can learn the meaning of these terms and how they are used in astronomy.

ASTRONOMY

V. Just For Fun

Finding galaxies!

❶   Observe more galaxies with your astronomy software. Find the following galaxies, and in the spaces provided, draw what you observe.

Whirlpool Galaxy

NGC 1427A

M 101

M 82

Bode's Galaxy

M 87

Sombrero Galaxy

Sunflower Galaxy

Hoag's Object

Cartwheel Galaxy

NGC 3314

❷   Look for other galaxies. Find the ones that you think are the most interesting, beautiful, intricate, or weird and record what you see.

ASTRONOMY

Galaxies!

More Galaxies!

ASTRONOMY

Even More Galaxies!

Experiment 20

Classifying Galaxies

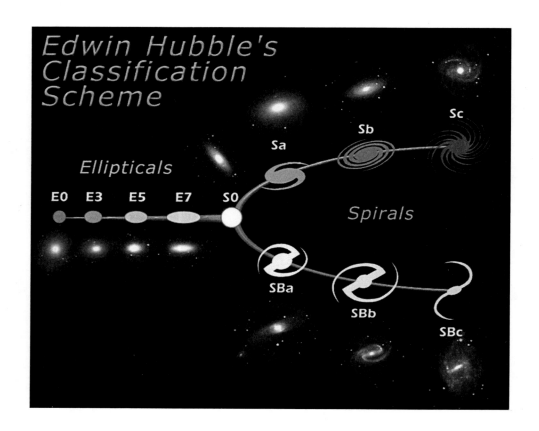

Introduction

By performing this citizen science experiment, you'll be helping astronomers gather data for their science project while you learn more about classifying galaxies.

I. Think About It

❶   Why do you think it is useful for astronomers to put galaxies into groups?

_____

_____

_____

_____

❷   Do you think it's easy for astronomers to group galaxies according to how they look? Why or why not?

_____

_____

_____

_____

❸   If you were looking at images of galaxies, what characteristics do you think you might observe that would help you sort them into groups?

_____

_____

_____

_____

ASTRONOMY

❹ Do you think astronomers have found all the different types of galaxies that exist? Why or why not?

_____

_____

_____

_____

_____

❺ What would you need to do to find out if a galaxy is a radio galaxy? Why?

_____

_____

_____

_____

_____

❻ If you discovered a new galaxy, what would you be most excited to find? Why?

_____

_____

_____

_____

_____

ASTRONOMY

II. Experiment 20: Classifying Galaxies   Date _____

Objective _____

_____

Hypothesis _____

_____

Materials

computer or tablet
internet connection

EXPERIMENT

With all the powerful land- and space-based telescopes in use today, astronomers have an unimaginably huge number of images of celestial bodies that need to be looked at. In this experiment you'll classify galaxies to help the astronomers who are studying them.

❶   Go to the Galaxy Zoo website: www.galaxyzoo.org/

❷   Under the "Profile" tab, create a Zooniverse account by entering a username and a password. In your user account you will be able to view images of the galaxies you've classified. Clicking on one of these images will bring up more information about it.

❸   Explore the website by looking at the information in the dropdown menus under the different tabs on the menu bar at the top. The "Story" page has information about the project and also has links to the websites for telescopes used to collect the images.

❹   Go to the "Classify" page. Click on the box that says "Examples." Reviewing the examples is important to help you understand how to classify the galaxies.

❺   Before starting to classify galaxies, go to the menu tab "Discuss" and click on "Talk." Here you will find images that have been classified by others and comments about the features they identified in an image. This information may help you in making your own classifications, but you may or may not agree with their conclusions.

❻   Start classifying! Click on the box that best answers the question about the galaxy image being displayed. To get an opposite view of the image (dark on light), click on the image or on the round blue button that says "Invert." This sometimes will reveal more features of the galaxy.

❼ Use the Results section or separate paper to keep notes as you go along.

Results

Galaxy Classification Notes	
Type	Features

ASTRONOMY

III. Conclusion

Based on your observations, how easy or difficult is it to classify galaxies? What did you discover about galaxies and images of them?

_____

_____

_____

_____

_____

_____

_____

_____

_____

_____

_____

_____

_____

_____

ASTRONOMY

IV. Why?

In this experiment you probably noticed that galaxies can be hard to classify. Some galaxies are so far away that the galaxy looks like a fuzzy little blob, making it impossible to tell whether it has any features. It can also be difficult to tell the orientation of a galaxy. Is the image a side view, a top view, or is the galaxy tilted? Spiral arms, galactic bulges, and bars can be hard to distinguish. All of these factors can affect galaxy classification when using the Hubble Tuning Fork method.

Astronomers have come up with additional ways to classify galaxies. One method is to observe their color. Studying data that have been collected shows that spiral galaxies tend to be more blue because they have more active star forming areas and so have younger, hotter, blue stars. On the other hand, elliptical galaxies seem to have little ongoing star formation and so are made of older, cooler, red stars, giving the ellipticals their characteristic reddish color. Like other classification methods, this one is not perfect, and not all galaxies can be classified accurately by using it.

You may be wondering why blue stars are hotter and red stars are cooler when it seems like the opposite should be true. Recall from *Exploring the Building Blocks of Science Book 4* that in the optical range of the electromagnetic spectrum, blue has shorter wavelengths than red. In general, shorter wavelengths are more energetic than longer wavelengths. For example, imagine you and a friend are each holding one end of a long jump rope. If you jiggle one end of the rope slowly, the waves created will have long wavelengths. If you put a lot of energy into pumping the end of the rope up and down, you'll create short wavelength, high frequency waves. In a star, vibration of atoms and molecules creates the wavelength and frequency of the electromagnetic waves emitted. Heat provides energy, so hotter stars will have more rapidly vibrating atoms and molecules and will emit more energetic, shorter, higher frequency blue wavelengths. Cooler stars have less energy and less rapidly moving atoms and molecules and will emit less energetic, longer, lower frequency red waves.

But why classify galaxies at all? Astronomers have lots of questions with no easy answers. How did/do galaxies form? How long does it take? Is there star formation in all galaxies? Why? How? Do galaxies morph from one type to another? If so—how, why, when? How do they interact with each other? How do they move through space? Were galaxies that existed billions of years ago formed in the same way as galaxies that are only 2 million years old? How can we tell how old are they are, anyway? Where did everything in the universe come from? What is dark matter? What is dark energy? Is there really dark matter and dark energy? Scientists expect that classifying and studying galaxies will help answer these and many other questions. Lots of theories may come and go during the discovery process.

ASTRONOMY

V. Just For Fun

❶ Go to Hubble Telescope website images: http://hubblesite.org/gallery/album/galaxy/

❷ Look at the images on the main "Galaxies" page (not the pages by class) and see if you can find one or more galaxies for each of the classes in the following chart.

❸ Click on the image to find out more about the galaxy. Fill in the chart with the name of the galaxy and interesting features. You can also print out your favorite images.

Galaxy Classifications

Spiral

Barred spiral

Galaxy Classifications

Elliptical

Lenticular

Irregular

Peculiar

Experiment 21

Searching for Nebulae

Courtesy of NASA, ESA and the Hubble SM4 ERO Team

Introduction

Use your online astronomy resource(s) to find nebulae.

I. Think About It

❶   How would you describe a nebula? Do you think there are different types of nebulae?

_____

_____

_____

_____

_____

❷   Would we be able to study the universe if a black hole were in our solar system? Why or why not?

_____

_____

_____

_____

_____

❸   Would we be able to study the universe if we lived on Jupiter? Why or why not?

_____

_____

_____

_____

ASTRONOMY

❹   How might our study of the universe change if Earth were located in the middle of a nebula?

_____

_____

_____

_____

_____

_____

❺   How many nebulae do you think exists in the universe?

_____

_____

_____

_____

_____

_____

❻   How do you think a nebula might change over time?

_____

_____

_____

_____

_____

ASTRONOMY

II. Experiment 21: Searching for Nebulae   Date_____

Objective _____

_____

Hypothesis _____

_____

Materials

computer
internet access

EXPERIMENT

❶   Open the online resource you chose from Experiment 18.

❷   Search for the following nebulae:

Helix Nebula
Crab Nebula
Cone Nebula
Cat's Eye Nebula
Eagle Nebula
Orion Nebula

❸   In the *Results* section make a drawing of what you observe for each nebula and note the location of each by naming the constellation it is in and/or any nearby constellations. Find out what type of nebula you're looking at, what features it has, and any other information you can discover about it. Record your observations.

Results

Helix Nebula

Location and Information

_____

_____

_____

_____

_____

_____

Crab Nebula

Location and Information

_____

_____

_____

_____

_____

_____

Cone Nebula

Location and Information

_____

_____

_____

_____

_____

_____

ASTRONOMY

Cat's Eye Nebula

Location and Information

_____

_____

_____

_____

_____

Eagle Nebula

Location and Information

_____

_____

_____

_____

_____

Orion Nebula

Location and Information

_____

_____

_____

_____

_____

_____

ASTRONOMY

III. Conclusion

What conclusions can you draw from your research?

_____

_____

_____

_____

_____

_____

_____

_____

_____

_____

_____

_____

_____

_____

_____

_____

_____

IV. Why?

In this experiment you used your online resource to identify a variety of nebulae with their different shapes and features. As you learned in the *Student Textbook,* nebulae are composed of clouds of dust and gas, with the gas being mostly hydrogen. Some nebulae contain areas where stars and planetary systems form.

One type of nebula you observed is the diffuse emission nebula. The Cone Nebula and Orion Nebula are examples of diffuse emission nebulae. A diffuse nebula is a cloud of thin, widespread gas and dust particles. If a diffuse nebula is big and massive enough, it can have star forming regions. When young, massive, hot stars are forming in a diffuse nebula, their high energy radiation causes the gas around them (mostly hydrogen gas) to emit light, making the nebula shine. In this way a diffuse nebula becomes a diffuse emission nebula.

The Eagle Nebula is also a diffuse emission nebula and provides some of the most striking images of any nebula. The "Pillars of Creation" region of the nebula is a large area of active star formation and has long protruding columns and other strange and beautiful shapes.

The Helix Nebula, Crab Nebula, and Cat's Eye Nebula are planetary nebulae. Recall that a planetary nebula is formed when clouds of gas and dust are ejected from a red giant star as it becomes a white dwarf.

In images we see of nebulae, they seem to be composed of dense areas of gases, dust, and stars, but in reality the tiny particles nebulae are made of are extremely far apart. Nebulae are not dense at all, but because they are so far away, we observe them as gaseous clouds with many different and often oddly shaped features.

Astronomers spend hours and even years looking for nebulae and other objects in the night sky. Some nebulae can be observed with the naked eye, but the detail that we see today in images of nebulae is possible because of modern telescopes and technology, including the imaging of electromagnetic wavelengths that are not visible to the unaided eye. As more advanced telescopes and new technologies become available, we may be able to find many more nebulae and observe some that are even more distant or even less dense than those we can view now. We will also learn more about the composition of nebulae and how they form and change.

ASTRONOMY

V. Just For Fun

Review the astronomy chapters in the *Student Textbook*. How many different types of objects described in the text can you find by using online resources? In the space provided, describe them in words and drawings.

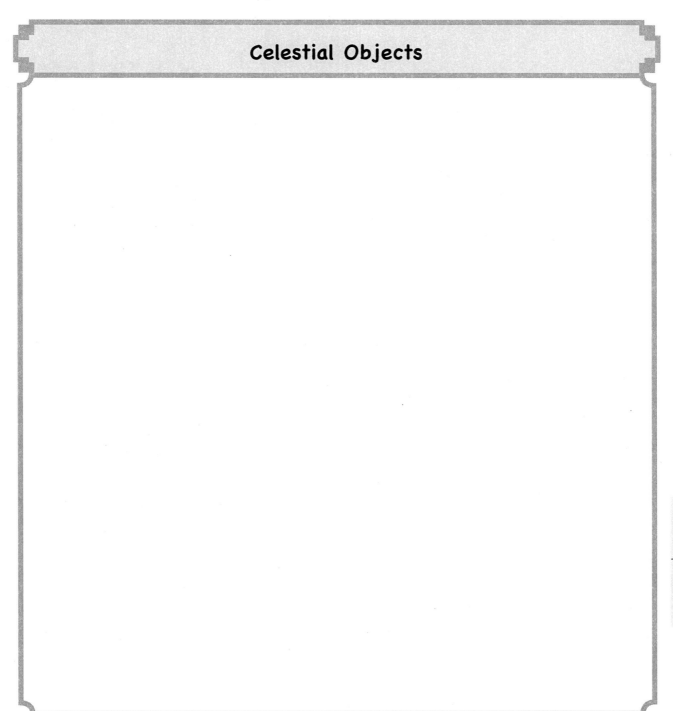

Celestial Objects

More Celestial Objects

Experiment 22

Reading a Research Article

Introduction

Discover how to look up and read a scientific research article.

I. Think About It

❶   Why do you think scientists publish research articles?

_____

_____

_____

_____

_____

❷   What do you think makes a research article good?

_____

_____

_____

_____

_____

❸   What sections do you think a research article would be divided into?

_____

_____

_____

_____

_____

❹ How easy or difficult do you think it would be for you to understand a research article?

_____

_____

_____

_____

_____

❺ How easy or difficult do you think it would be for the general public to understand a research article?

_____

_____

_____

_____

_____

❻ Do you think it is important for scientists to write research articles? Why or why not?

_____

_____

_____

_____

_____

II. Experiment 22: Reading a Research Article   Date _____

This experiment does not require an objective nor a hypothesis.

Materials

computer
internet access

EXPERIMENT

❶ In the *Student Textbook* you read about many different subjects. Review the text and pick a few subjects that you would be interested in exploring further. List them below.

_____

_____

_____

_____

_____

_____

❷ From the list above, select one topic that you would most like to learn more about. Using a computer, do a search for "scholarly articles" for this topic. The search will likely result in a large number of articles. Scroll through the first ten and then list the titles of the three that sound most interesting.

_____

_____

_____

_____

_____

❸   Choose one of the three papers you listed in Step ❷ and look at it online. If a purchase is required to read the paper, look for a different paper that is free.

❹   Read the abstract of the paper you have chosen. Note below whether you can understand the abstract, parts of the abstract, or hardly any of it. Describe the parts you can understand and also any difficulties you may have with understanding the article.

_____

_____

_____

_____

_____

_____

_____

❺   Look through the rest of the paper and note how it is structured. Write the titles of the major sections below. These will include an introduction, methods and materials, results, and the conclusion.

_____

_____

_____

_____

_____

_____

❻   Read the introduction. Write a brief summary of what you think it is saying.

_____

_____

_____

_____

_____

_____

_____

❼   Read the materials and methods section and the discussion section, if there is one, and note below how well you understand them. Write a brief summary of the parts you understand.

_____

_____

_____

_____

_____

_____

❽   Read the conclusion section of the paper, if it has one, and note how well you undersand it. Write a brief summary of what you understand.

_____

_____

_____

_____

_____

_____

_____

❾   Look at the references that are listed in the article and note how the references correspond to different sections of the paper. Reread a few sections that have reference notations and observe the title of each reference. Record below the references that sound most interesting and why. Also note how you think looking up some of the specific papers that are referenced could help you to dig deeper into the topic.

_____

_____

_____

_____

_____

_____

_____

III. Conclusion

What conclusions can you draw from your research?

_____

_____

_____

_____

_____

_____

_____

_____

_____

_____

_____

_____

_____

_____

_____

IV. Why?

Learning how to read a scientific paper is an important skill and necessary for all scientists. Most scientific papers are presented in a common format starting with an abstract followed by the introduction, materials and methods, results, sometimes a discussion section, and finally a section for conclusions. All scientific papers reference other people's work and list references at the end of the paper.

The abstract is usually a brief summary of background information, a description of the work performed, and the conclusions drawn from the research. In the introduction the paper will go into more detail about why the work was done; the materials and methods section describes how the work was done; and the discussion and conclusion sections talk about what the work may mean. To get a quick idea of whether or not a paper might be interesting, many scientists will read the abstract and then scan the materials and conclusions sections. Many scientists will also flip to the references to determine who the authors consider important resources. Many times the references listed in an article are more interesting and more valuable than the article itself.

If the article is interesting enough, a scientist will likely read it in detail, paying close attention to the materials and methods section. If they do work similar to the authors of the paper, they are likely to critique the methods section and evalutate the data in the results section. This is a great way for scientists to learn about how other labs do research and the data other scientists are collecting.

When a paper is submitted for publication in a scientific journal, scientists who are doing similar work are often called on to be "reviewers." These reviewers read the paper while focusing mostly on the methods and results. If the methods or results are weak or don't convincingly support the conclusion the author presents, the paper is likely to be rejected. For this reason, good scientific papers can take years and years of data collection and careful analysis.

V. Just For Fun

Write your own scholarly scientific article!

Pick one of the experiments in this workbook and write your completed experiment as a scientific article. Include a title and sections for the abstract, introduction, materials and methods, and conclusion.

Title _____

_____

p. 2

p. 3

Appendix: Globular Clusters

On the following pages are several data tables with information about globular clusters observed in the Milky Way. Learning how to read data tables and sorting through the information to find the data needed for an experiment is an important part of scientific investigation.

Key

M, NGC/IC, ID/Name/Crossref

Messier, NGC, or IC number and other identification or name

Con

Constellation name (number of globular clusters in parentheses)

RA, Dec (2000)

Right Ascension and Declination for epoch 2000.0

R_Sun, R_gc

Distance from our Sun and the Galactic Center in thousands of light years (kly)

m_v

Apparent visual magnitude

dim

Apparent dimension in arc minutes

APPENDIX

M	NGC/IC	ID/Name/Crossref	Con	RA (2000)	DEC	R_Sun	R_gc	m_v	dim
	104	47 Tuc Lac I.1	Tucana (2)	00:24:05.2	-72:04:51	14.7	24.1	3.95	50.0
	288	H 6.20	Scl	00:52:47.5	-26:35:24	28.7	39.1	8.09	13.0
	362	Dun 62	Tucana (2)	01:03:14.3	-70:50:54	27.7	30.3	6.40	14.0
		Whiting 1	Cet	02:02:56.8	-03:15:10				1.2
	1261	Dun 337	Horologium (2)	03:12:15.3	-55:13:01	53.5	59.4	8.29	6.6
		Pal 1	Cep	03:33:23.0	+79:34:50	35.6	55.4	13.18	2.8
		AM 1 E 1	Horologium (2)	03:55:02.7	-49:36:52	397.6	401.8	15.72	0.5
		Eri	Eri	04:24:44.5	-21:11:13	294.2	310.5	14.70	1.0
		Pal 2	Aur	04:46:05.9	+31:22:51	90.0	115.5	13.04	2.2
	1851	Dun 508	Col	05:14:06.3	-40:02:50	39.5	54.5	7.14	12.0
M 79	1904		Lep	05:24:10.6	-24:31:27	42.1	61.3	7.73	9.6
	2298	Dun 578	Pup	06:48:59.2	-36:00:19	34.9	51.2	9.29	5.0
	2419	H 1.218	Lyn	07:38:08.5	+38:52:55	274.6	298.4	10.39	4.6
		Koposov 2	Gem	07:58:17.0	+26:15:18	130			
		Pyxisis	Pyx	09:07:57.8	-37:13:17	129.4	135.9	12.90	4.0
	2808	Dun 265	Car	09:12:02.6	-64:51:47	31.2	36.2	6.20	14.0
		E 3	Cha	09:20:59.3	-77:16:57	14.0	24.8	11.35	10:
		Pal 3	Sex	10:05:31.4	+00:04:17	302.3	312.8	14.26	1.6
		Segue 1	Leo	10:07:04	+12:47:30	75.0	14.7	4.5	
	3201	Dun 445	Vel	10:17:36.8	-46:24:40	16.3	29.0	6.75	20.0
		Pal 4	UMa	11:29:16.8	+28:58:25	356.2	364.6	14.20	1.3
		Koposov 1	Virgo (2)	11:59:18.5	+12:15:36	160			
	4147	H 1.19	Coma Berenices (3)	12:10:06.2	+18:32:31	62.9	69.5	10.32	4.4
	4372		Mus	12:25:45.4	-72:39:33	18.9	23.2	7.24	5.0
		Rup 106	Cen	12:38:40.2	-51:09:01	69.1	60.3	10.90	2.0
M 68	4590		Hya	12:39:28.0	-26:44:34	33.3	32.9	7.84	11.0
	4833	Lac I.4 Dun 164	Mus	12:59:35.0	-70:52:29	21.2	22.8	6.91	14.0
M 53	5024		Coma Berenices (3)	13:12:55.3	+18:10:09	58.0	59.6	7.61	13.0
	5053	H 6.7	Coma Berenices (3)	13:16:27.0	+17:41:53	53.5	55.1	9.47	10.0
	5139	Omega Cen Lac I.5	Cen	13:26:45.9	-47:28:37	17.3	20.9	3.68	55.0
M 3	5272		CVn	13:42:11.2	+28:22:32	33.9	39.8	6.19	18.0
	5286	Dun 388	Cen	13:46:26.5	-51:22:24	35.9	27.4	7.34	11.0
		AM 4	Hya	13:56:21.2	-27:10:04	97.5	83.2	15.90	3.0
	5466	H 6.9	Boo	14:05:27.3	+28:32:04	51.8	52.8	9.04	9.0
	5634	H 1.70	Virgo (2)	14:29:37.3	-05:58:35	82.2	69.1	9.47	5.5
	5694	H 2.196	Hya	14:39:36.5	-26:32:18	113.2	94.9	10.17	4.3

	NGC	Other	Const.	RA	Dec				
	5824	I4499	Aps	15:00:18.5	-82:12:49	61.6	51.2	9.76	8.0
			Lup	15:03:58.5	-33:04:04	104.4	84.1	9.09	7.4
		Pal 5	SerCp	15:16:05.3	-00:06:41	75.7	60.7	11.75	8.0
	5897	H 6.8 H 6.19	Lib	15:17:24.5	-21:00:37	40.4	23.8	8.53	11.0
M 5	5904		SerCp	15:18:33.8	+02:04:58	24.5	20.2	5.65	23.0
	5927	Dun 389	Lup	15:28:00.5	-50:40:22	24.8	14.7	8.01	6.0
	5946		Norma (3)	15:35:28.5	-50:39:34	34.6	18.9	9.61	3.0
		BH 176	Norma (3)	15:39:07.3	-50:03:02	50.9	31.6	14.00	3.0
	5986	Dun 552	Lup	15:46:03.5	-37:47:10	33.9	15.7	7.52	9.6
		Lynga 7	Norma (3)	16:11:03.0	-55:18:52	23.5	2.5		
		Pal 14 AvdB	Her	16:11:04.9	+14:57:29	241.0	225.0	14.74	2.5
M 80	6093		Scorpio (20)	16:17:02.5	-22:58:30	32.6	12.4	7.33	10.0
M 4	6121	Lac I.9	Scorpio (20)	16:23:35.5	-26:31:31	7.2	19.2	5.63	36.0
	6101	Dun 68	Aps	16:25:48.6	-72:12:06	49.9	36.2	9.16	5.0
	6144	H 6.10	Scorpio (20)	16:27:14.1	-26:01:29	27.7	8.5	9.01	7.4
	6139	Dun 536	Scorpio (20)	16:27:40.4	-38:50:56	32.9	11.7	8.99	8.2
		Terzan 3	Scorpio (20)	16:28:40.1	-35:21:13	24.5	7.8	12.00	3.0
M 107	6171	H 6.40	Ophiuchus (25)	16:32:31.9	-13:03:13	20.9	10.8	7.93	13.0
		1636-283 ESO452-SC11	Scorpio (20)	16:39:25.5	-28:23:52	25.4	6.5	12.00	1.2
M 13	6205		Her	16:41:41.5	+36:27:37	25.1	28.4	5.78	20.0
	6229	H 4.50	Her	16:46:58.9	+47:31:40	99.1	96.8	9.39	4.5
M 12	6218		Ophiuchus (25)	16:47:14.5	-01:56:52	16.0	14.7	6.70	16.0
		FSR 1735 2MASS-GC03	Arae (5)	16:52:10.6	-47:03:29	29.7	10.4	0.8	
	6235	H 2.584	Ophiuchus (25)	16:53:25.4	-22:10:38	37.2	13.4	9.97	5.0
M 10	6254		Ophiuchus (25)	16:57:08.9	-04:05:58	14.4	15.0	6.60	20.0
	6256		Scorpio (20)	16:59:32.6	-37:07:17	27.4	5.9	11.29	4.1
		Pal 15	Ophiuchus (25)	17:00:02.4	-00:32:31	145.5	123.6	14.00	3.0
M 62	6266		Ophiuchus (25)	17:01:12.6	-30:06:44	22.5	5.5	6.45	15.0
M 19	6273		Ophiuchus (25)	17:02:37.7	-26:16:05	28.0	5.2	6.77	17.0
	6284	H 6.11	Ophiuchus (25)	17:04:28.8	-24:45:53	49.9	24.8	8.83	6.2
	6287	H 2.195	Ophiuchus (25)	17:05:09.4	-22:42:29	30.3	6.8	9.35	4.8
	6293	H 6.12	Ophiuchus (25)	17:10:10.4	-26:34:54	28.7	4.6	8.22	8.2
	6304	H 1.147	Ophiuchus (25)	17:14:32.5	-29:27:44	19.6	7.2	8.22	8.0
	6316	H 1.45	Ophiuchus (25)	17:16:37.4	-28:08:24	35.9	10.4	8.43	5.4
M 92	6341		Her	17:17:07.3	+43:08:11	26.7	31.3	6.44	14.0
	6325		Ophiuchus (25)	17:17:59.2	-23:45:57	26.1	3.6	10.33	4.1
M 9	6333		Ophiuchus (25)	17:19:11.8	-18:30:59	25.8	5.5	7.72	12.0
	6342	H 1.149	Ophiuchus (25)	17:21:10.2	-19:35:14	28.0	5.5	9.66	4.4
	6356	H 1.48	Ophiuchus (25)	17:23:35.0	-17:48:47	49.6	24.8	8.25	10.0

	6355	H 1.46	Ophiuchus (25)	17:23:58.6	-26:21:13	31.0	5.9	9.14	4.2
	6352	Dun 417	Arae (5)	17:25:29.2	-48:25:22	18.6	10.8	7.96	9.0
		I1257	Ophiuchus (25)	17:27:08.5	-07:05:35	81.5	58.4	13.10	5.0
	6366	Terzan 2 HP 3	Scorpio (20)	17:27:33.4	-30:48:08	28.4	2.9	14.29	0.6
			Ophiuchus (25)	17:27:44.3	-05:04:36	11.7	16.3	9.20	13.0
		Terzan 4 HP 4	Scorpio (20)	17:30:38.9	-31:35:44	29.7	4.2	16.00	0.7
		HP 1 BH 229	Ophiuchus (25)	17:31:05.2	-29:58:54	46.0	19.9	11.59	1.2
	6362	Dun 225	Arae (5)	17:31:54.8	-67:02:53	24.8	16.6	7.73	15.0
		Liller 1	Scorpio (20)	17:33:24.5	-33:23:20	34.2	8.5	16.77	12.6
	6380	Ton 1	Scorpio (20)	17:34:28.0	-39:04:09	34.9	10.4	11.31	3.6
		FSR 1767	Scorpio (20)	17:35:43	-36:21:28	4.9	18.6		
		Terzan 1 HP 2	Scorpio (20)	17:35:47.8	-30:28:11	18.3	8.2	15.90	2.4
		Ton 2 Pismis 26	Scorpio (20)	17:36:10.5	-38:33:12	26.4	4.6	12.24	2.2
	6388	Dun 457	Scorpio (20)	17:36:17.0	-44:44:06	32.6	10.4	6.72	10.4
M 14	6402		Ophiuchus (25)	17:37:36.1	-03:14:45	30.3	13.4	7.59	11.0
	6401	H 1.44	Ophiuchus (25)	17:38:36.9	-23:54:32	34.2	8.8	9.45	4.8
	6397	Lac III.11 Dun 366	Arae (5)	17:40:41.3	-53:40:25	7.5	19.6	5.73	31.0
		Pal 6	Ophiuchus (25)	17:43:42.2	-26:13:21	19.2	7.2	11.55	1.2
	6426	H 2.587	Ophiuchus (25)	17:44:54.7	+03:10:13	67.5	47.6	11.01	4.2
		Djorg 1	Scorpio (20)	17:47:28.3	-33:03:56	39.1	13.4	13.60	
		Terzan 5 Terzan 11	Sagittarius (34)	17:48:04.9	-24:48:45	33.6	7.8	13.85	2.4
	6440	H 1.150	Sagittarius (34)	17:48:52.6	-20:21:34	27.4	4.2	9.20	4.4
	6441	Dun 557	Scorpio (20)	17:50:12.9	-37:03:04	38.1	12.7	7.15	9.6
		Terzan 6 HP 5	Scorpio (20)	17:50:46.4	-31:16:31	31.0	5.2	13.85	1.4
	6453		Scorpio (20)	17:50:51.8	-34:35:55	31.3	5.9 10.0	7.6	
		UKS 1 UKS 1751-241	Sagittarius (34)	17:54:27.2	-24:08:43	27.1	2.6	17.29	2.0
	6496	Dun 460	Scorpio (20)	17:59:02.0	-44:15:54	37.5	14.0	8.54	5.6
		Terzan 9	Sagittarius (34)	18:01:38.8	-26:50:23	21.2	5.2	16.00	0.2
		Djorg 2 E456-SC38	Sagittarius (34)	18:01:49.1	-27:49:33	21.9	4.6	9.90	9.9
	6517	H 2.199	Ophiuchus (25)	18:01:50.6	-08:57:32	35.2	14.0	10.23	4.0
		Terzan 10	Sagittarius (34)	18:02:57.4	-26:04:00	18.6	7.8	14.90	1.5
	6522	H 1.49	Sagittarius (34)	18:03:34.1	-30:02:02	25.4	2.0	8.27	9.4
	6535		SerCd	18:03:50.7	-00:17:49	22.2	12.7	10.47	3.4
	6528	H 2.200	Sagittarius (34)	18:04:49.6	-30:03:21	25.8	2.0	9.60	5.0
	6539		SerCd	18:04:49.8	-07:35:09	27.4	10.1	9.33	7.9
	6540	H 2.198 Djorg 3	Sagittarius (34)	18:06:08.6	-27:45:55	12.1	14.4	9.30	1.5
	6544	H 2.197	Sagittarius (34)	18:07:20.6	-24:59:51	8.8	17.3	7.77	9.2
	6541	Dun 473	CrA	18:08:02.2	-43:42:20	22.8	7.2	6.30	15.0
		2MASS-GC01	Sagittarius (34)	18:08:21.8	-19:49:47	11.7	14.7	3.3	

M	NGC	Name	Constellation	RA	Dec				
		ESO 280-SC06	Arae (5)	18:09:06	-46:25:24	70.7	46.6	1.5	
	6553	H 4.12	Sagittarius (34)	18:09:15.6	-25:54:28	19.6	7.2	8.06	9.2
		2MASS-GC02	Sagittarius (34)	18:09:36.5	-20:46:44	13.0	13.4	1.9	4.2
	6558		Sagittarius (34)	18:10:18.4	-31:45:49	24.1	3.3	9.26	8.0
		I1276 Pal 7	SerCd	18:10:44.2	-07:12:27	17.6	12.1	10.34	1.0
		Terzan 12	Sagittarius (34)	18:12:15.8	-22:44:31	15.7	11.1	15.63	6.4
	6569	H 2.201 Dun 619	Sagittarius (34)	18:13:38.9	-31:49:35	34.9	9.5	8.55	1.3
		AL 3	Sagittarius (34)	18:14:05.7	-28:38:08	43.7	22.8	8.27	6.6
	6584	Dun 376	Tel	18:18:37.7	-52:12:54	25.8	3.9	7.87	8.8
	6624	H 1.50	Sagittarius (34)	18:23:40.5	-30:21:40				
M 28	6626	Lac I.11	Sagittarius (34)	18:24:32.9	-24:52:12	18.3	8.8	6.79	11.2
	6638	H 1.51	Sagittarius (34)	18:30:56.2	-25:29:47	31.2	7.5	9.02	7.3 M, 6
	6637	Lac I.12 Dun 613	Sagittarius (34)	18:31:23.2	-32:20:53	29.7	6.2	7.64	9.8
	6642	H 2.205	Sagittarius (34)	18:31:54.3	-23:28:35	27.4	5.5	9.13	5.8
	6652		Sagittarius (34)	18:35:45.7	-32:59:25	32.9	9.1	8.62	6.0
M 22	6656		Sagittarius (34)	18:36:24.2	-23:54:12	10.4	16.0	5.10	32.0
		Pal 8	Sagittarius (34)	18:41:29.9	-19:49:33	42.1	18.3	11.02	5.2
M 70	6681	Dun 614	Sagittarius (34)	18:43:12.7	-32:17:31	29.4	6.8	7.87	8.0
		GLIMPSE-C01	Aquila (4)	18:48:49.7	-01:29:50			10-17	
	6712	H 1.47	Sct	18:53:04.3	-08:42:22	22.5	11.4	8.10	9.8
M 54	6715	Dun 624	Sagittarius (34)	18:55:03.3	-30:28:42	87.3	62.6	7.60	12.0
	6717	H 3.143 Pal 9	Sagittarius (34)	18:55:06.2	-22:42:03	23.1	7.8	9.28	5.4
	6723	Dun 573	Sagittarius (34)	18:59:33.2	-36:37:54	28.4	8.4	7.01	13.0
	6749	Berkeley 42	Aquila (4)	19:05:15.3	+01:54:03	25.8	16.3	12.44	4.0
	6752	Dun 295	Pav	19:10:51.8	-59:58:55	13.0	17.0	5.40	29.0
	6760		Aquila (4)	19:11:12.1	+01:01:50	24.1	15.7	8.88	9.6
M 56	6779		Lyr	19:16:35.5	+30:11:05	32.9	31.6	8.27	8.8
		Terzan 7	Sagittarius (34)	19:17:43.7	-34:39:27	75.7	52.2	12.00	1.2
		Pal 10	Sge	19:18:02.1	+18:34:18	19.2	20.9	13.22	4.0
		Arp 2	Sagittarius (34)	19:28:44.1	-30:21:14	93.3	69.8	12.30	2.3
M 55	6809	Lac I.14 Dun 620	Sagittarius (34)	19:39:59.4	-30:57:44	17.3	12.7	6.32	19.0
		Terzan 8	Sagittarius (34)	19:41:45.0	-34:00:01	84.8	62.3	12.40	3.5
		Pal 11	Aquila (4)	19:45:14.4	-08:00:26	42.4	25.8	9.80	10.0
M 71	6838		Sge	19:53:46.1	+18:46:42	13.0	21.9	8.19	7.2
M 75	6864	H 1.103	Sagittarius (34)	20:06:04.8	-21:55:17	67.5	47.6	8.52	6.8
	6934		Delphinus (2)	20:34:11.6	+07:24:15	51.2	41.7	8.83	7.1
M 72	6981		Aqr	20:53:27.9	-12:32:13	55.4	42.1	9.27	6.6
	7006	H 1.52	Delphinus (2)	21:01:29.5	+16:11:15	135.4	126.5	10.56	3.6
M 15	7078		Peg	21:29:58.3	+12:10:01	33.6	33.9	6.20	18.0

M	NGC	Name	Constellation	RA	Dec				
M 2	7089		Aqr	21:33:29.3	-00:49:23	37.5	33.9	6.47	16.0
M 30	7099		Cap	21:40:22.0	-23:10:45	26.1	23.2	7.19	12.0
		Pal 12 Cap Dwarf	Cap	21:46:38.8	-21:15:03	62.3	51.9	11.99	2.9
		Pal 13	Peg	23:06:44.4	+12:46:19	84.1	87.0	13.80	0.7
	7492	H 3.558	Aqr	23:08:26.7	-15:36:41	84.1	81.2	11.29	4.2

References

V. Berokurov, D. B. Zucker, N. W. Evans, J. T. Kleyna, S. Koposov, S. T. Hodgkin, M. J. Irwin, G. Gilmore, M. I. Wilkinson, M. Fellhauer, D. M. Bramich, P. C. Hewett, S. Vidrir, J. T. A. de Jong, J. A. Smith, H.-W. Rix, E. F. Bell, R. F. G. Wyse, H. J. Newberg, P. A. Mayeur, B. Yanny, C. M. Rockosi, O. Y. Gnedin, D. P. Schneider, T. C. Beers, J. C. Barentine, H. Brewington, J. Brinkmann, M. Harvanek, S. J. Kleinman, J. Krzesinski, D. Long, A. Nitta, S. A. Sneddon, 2007. Cats and Dogs, Hair and a Hero: A Quintet of New Milky Way Companions. Astrophysical Journal, Vol. 654, Issue 2, pp. 897-906 (January 2007) [ADS: 2007ApJ...654..897B] - [Preprint: astro-ph/0608448] Discovery announce of Segue 1.

C. Bonatto, E. Bica, S. Ortolani and B. Barbuy, 2007. FSR 1767 - a new globular cluster in the Galaxy. To be published in: Monthly Notices of the Royal Astronomical Society. [Preprint: arXiv:0708.0501[astro-ph]] Discovery announce of FSR-1767.

D. Froebrich, H. Meusinger and A. Scholz, 2007. SR 1735 - A new globular cluster candidate in the inner Galaxy. To appear in: Monthly Notices of the Royal Astronomical Society (2007). [Preprint: astro-ph/0703318] Discovery announce of FSR-1735.

W. E. Harris, 1996-1999. Catalog of Parameters for Milky Way Globular Clusters. AJ, 112, 1487. Revision of June 22, 1999. Available online; also see references and the potentially more current original site.

R. J. Hurt, et al., 2000. Serendipitous 2MASS Discoveries near the Galactic Plane: A Spiral Galaxy and Two Globular Clusters. The Astronomical Journal, Volume 120, Issue 4, pp. 1876-1883 (10/2000). [ADS: 2000AJ....120.1876H] - [Preprint] Discovery announce of two new globulars, 2MASS-GC01 and 2MASS-GC02.

Henry Kobulnicky, B. L. Babler, T. M. Bania, R. A. Benjamin, B. A. Buckalew, R. Canterna, E. Churchwell, D. Clemens, M. Cohen, J. M. Darnel, J. M. Dickey, R. Indebetouw, J. M. Jackson, A. Kutyrev, A. P. Marston, J. S. Mathis, M. R. Meade, E. P. Mercer, A. J. Monson, J. P. Norris, M. J. Pierce, R. Shah, J. R. Stauffer, S. R. Stolovy, B. Uzpen, C. Watson, B. A. Whitney, M. J. Wolff, and M. G. Wolfire, 2004. Newfound Star Cluster may be final Milky Way 'Fossil.' Spitzer Science Center News Release 2004-16. Discovery announce of the new globular GLIMPSE-C01.

S. Koposov, J. T. A. de Jong, H.-W. Rix, D. B. Zucker, N. W. Evans, G. Gilmore, M. J. Irwin, E. F. Bell, 2007. The discovery of two extremely low luminosity Milky Way globular clusters. Submitted to Astrophysical Journal. [Preprint: arXiv:0706.0019[astro-ph]] Discovery paper of Koposov 1 and Koposov 2.

S. Ortolani, E. Bica and B. Barbuy (2000). ESO 280-SC06: a new globular cluster in the Galaxy. Astronomy and Astrophysics, Vol. 361, pp. L57-L59 (September 2000). [ADS: 2000A&A...361L..57O] Discovery announce of the new globular, ESO 280-SC06.

S. Ortolani, E. Bica and B. Barbuy, 2006. AL 3 (BH 261): a new globular cluster in the Galaxy. Astrophysical Journal, Vol. 646, Issue 2, pp. L115-L118 (August 2006) [ADS: 2006ApJ...646L.115O] - [Preprint: astro-ph/0606718]. Discovery announce of the globular nature of AL-3.

More REAL SCIENCE-4-KIDS Books
by Rebecca W. Keller, PhD

Building Blocks Series yearlong study program — each Student Textbook has accompanying Laboratory Notebook, Teacher's Manual, Lesson Plan, Study Notebook, Quizzes, and Graphics Package

Exploring Science Book K (Activity Book)
Exploring Science Book 1
Exploring Science Book 2
Exploring Science Book 3
Exploring Science Book 4
Exploring Science Book 5
Exploring Science Book 6
Exploring Science Book 7
Exploring Science Book 8

Focus On Series unit study program — each title has a Student Textbook with accompanying Laboratory Notebook, Teacher's Manual, Lesson Plan, Study Notebook, Quizzes, and Graphics Package

Focus On Elementary Chemistry
Focus On Elementary Biology
Focus On Elementary Physics
Focus On Elementary Geology
Focus On Elementary Astronomy

Focus On Middle School Chemistry
Focus On Middle School Biology
Focus On Middle School Physics
Focus On Middle School Geology
Focus On Middle School Astronomy

Focus On High School Chemistry

Super Simple Science Experiments

21 Super Simple Chemistry Experiments
21 Super Simple Biology Experiments
21 Super Simple Physics Experiments
21 Super Simple Geology Experiments
21 Super Simple Astronomy Experiments
101 Super Simple Science Experiments

Note: A few titles may still be in production.

Gravitas Publications Inc.
www.gravitaspublications.com
www.realscience4kids.com